DISCOVERING WAVELETS

DISCOVERING WAVELETS

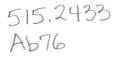

Edward Aboufadel and Steven Schlicker

A Wiley-Interscience Publication

JOHN WILEY & SONS, INC.

New York • Chichester • Weinheim • Brisbane • Singapore • Toronto

Copyright © 1999 by John Wiley & Sons, Inc. All rights reserved.

Published simultaneously in Canada.

No part of this publication may be reproduced, stored in a retrieval system or transmitted in any form or by any means, electronic, mechanical, photocopying, recording, scanning or otherwise, except as permitted under Section 107 or 108 of the 1976 United States Copyright Act, without either the prior written permission of the Publisher, or authorization through payment of the appropriate per-copy fee to the Copyright Clearance Center, 222 Rosewood Drive, Danvers, MA 01923, (978) 750-8400, fax (978) 750-4744. Requests to the Publisher for permission should be addressed to the Permissions Department, John Wiley & Sons, Inc., 605 Third Avenue, New York, NY 10158-0012, (212) 850-6011, fax (212) 850-6008, E-Mail: PERMREQ @ WILEY.COM.

For ordering and customer service, call 1-800-CALL-WILEY.

Library of Congress Cataloging in Publication Data:

Aboufadel, Edward, 1965–
    Discovering wavelets / Edward Aboufadel and Steven Schlicker.
        p.   cm.
    Includes bibliographical references.
    ISBN 0-471-33193-7 (alk. paper)
    1. Wavelets (Mathematics).   I. Schlicker, Steven, 1958–   .
    II. Title.
    QA403.3.A34   1999
    515'.2433—dc21   99-31029
                                                      CIP

Printed in the United States of America
10 9 8 7 6 5 4 3 2

Preface

For the past several years, reflecting the excitement and creativity surrounding the subject of wavelets, articles about wavelets have appeared in professional publications [4, 17, 20, 23, 24, 22, 26, 33, 35, 41], on the World Wide Web [16, 39, 27], and in mainstream magazines and newspapers [9, 31]. Much of this enthusiasm for wavelets comes from known and from potential applications. For example, wavelets have found use in image processing, in the restoration of recordings, and in seismology [3, 9, 23, 24, 30].

Many books are available on wavelets [10, 19, 21, 25, 30, 42], but most are written at such a level that only research mathematicians can read them. The purpose of this book is to make wavelets accessible to anyone with a background in basic linear algebra (for example, graduate and undergraduate students), and to serve as an introduction for the nonspecialist. The level of the applications and the format of this book make it an excellent textbook for an introductory course on wavelets or as a supplement to a first or second course in linear algebra or numerical analysis. Potential readers would be intrigued by the discussion of the Wavelet/Scalar Quantization Standard, currently used by the Federal Bureau of Investigation to compress, transmit, and store images of fingerprints [3]. In Britain, Scotland Yard also uses wavelets for the same purpose [4].

The projects that are contained within this book allow real applications to be incorporated into the mathematics curriculum. This fits well with the current trend of infusing mathematics courses with applications; an approach which is summarized well by Avner Friedman and John Lavery in their statement about an industrial mathematics program:

"[this approach provides] students an immense opportunity for greater and deeper contributions in all areas of the natural and social sciences, engineering and technology." [15]

The practice of either basing a mathematics course on applications [1] or infusing a course with applications has been seen primarily in calculus courses. However, this enthusiasm for real applications has not been as obvious in our upper-level mathematics courses, such as linear algebra, abstract algebra, or number theory, which are taken primarily by mathematics majors. Many of these majors intend to be teachers, and applications such as wavelets can provide breadth to the curriculum. As stated in a recent report of the Mathematical Association of America:

"it is vitally important that [prospective math teachers'] undergraduate experience provide a broad view of the discipline, [since] for the many students who may never make professional use of mathematics, depth through breadth offers a strong base for appreciating the true power and scope of the mathematical sciences." [32]

The major benefit of this book is that it presents basic and advanced concepts of wavelets in a way that is accessible to anyone with only a basic knowledge of linear algebra. The discussion of wavelets begins with an idea of Gil Strang [33] to represent the Haar wavelets as vectors in $\mathbb{R}^n$, and is driven by the desire to introduce readers to how the FBI is compressing fingerprint images. Chapter 1 introduces the basic concepts of wavelet theory in a concrete setting: the Haar wavelets along with the problem of digitizing fingerprints. The rest of the book builds on this material. To fully understand the concepts in this chapter, a reader need only have an understanding of basic linear algebra ideas — matrix multiplication, adding and multiplying vectors by scalars, linear independence and dependence.

Chapter 2 builds on the ideas presented in chapter 1, developing more of the theory of wavelets along with function spaces. Readers get a better sense of how one might deduce the ideas presented in chapter 1. To fully understand the material in this chapter, a reader needs more sophisticated mathematics, such as inner product spaces and projections. The necessary background material from linear algebra that a reader needs to know to fully understand the discussion of wavelets in this book is contained in appendix A.

Chapter 3 features more advanced topics such as filters, multiresolution analysis, Daubechies wavelets, and further applications. These topics are all introduced by comparison to the material developed with the Haar wavelets in chapters 1 and 2. These topics would be of interest to anyone who desires to read some of the more technical books or papers on wavelets, or to anyone seeking a starting point for research projects. There are some new concepts introduced in this chapter (e.g. density, fixed-point algorithms, and L^2 spaces). They are discussed in enough detail to allow the reader to understand these concepts and how they relate to wavelets, but not in so much detail that these ancillary topics distract from the major topic of wavelets. For example, the Fourier transform does not appear until the final section of chapter 3.

Chapter 4 contains projects that could be used in linear algebra courses. Along with these projects, some of the problems introduce advanced topics that could be used as starting points for research by undergraduates. There are also appendices that review linear algebra topics and present *Maple* commands that are useful for some of the problems.

These notes originated in courses (Linear Algebra II) that the authors taught during summers 1996 and 1997 at Grand Valley State University in Allendale, Michigan. Students in each of these courses had previously completed Linear Algebra I, where they learned to solve systems of linear equations, were introduced to matrix and vector operations, and encountered for the first time the fundamental ideas of spanning sets, linear independence and dependence, bases, and dimension. The topics for the second semester course were grouped around two major themes: linear transformations and orthogonal projections. In the 1996 course, about one of every five class meetings was set aside to learn about wavelets, and the material in these notes was timed to coincide with the flow of topics in the rest of the course. (For example, using orthogonal projections to approximate functions with wavelets was done as we studied orthogonal projections in inner product spaces.) Along with the information presented in these notes, various articles [2, 3, 9, 16, 23, 22, 31] were distributed to students as required readings.

In the 1997 course, groups of students submitted written reports based on a subset of the problems in chapters 1 and 2. In addition, each group created a gray-scale image in a 16-by-16 grid of pixels using a program, *Pixel Images*, written by Schlicker. Each group processed, compressed, and decompressed their image using Haar wavelets and entropy coding. A modified version of this activity is included in chapter 4.

While we initially introduced wavelets into the second semester linear algebra course, many of the tools and concepts can fit equally well in a first semester linear algebra course. In the last two years we have used various approaches in both semesters of linear algebra to expose our students to this valuable and exciting area of mathematics. In all of these courses, the students have expressed that they appreciated seeing a connection between what they were learning in college and what they saw happening in the world.

A note on how to use this book. To learn mathematics it is important to become conversant with the terminology and to actually work some problems. Throughout this book, key terms and phrases are highlighted in italics. For pedagogical purposes, we have included some terminology that is not standard in the literature: the phrase *daughter wavelets* in chapter 1 is used to describe the functions that are obtained from dilations and translations of the mother wavelet, *son wavelets* in chapter 2 describes the functions that are obtained from dilations and translations of the father wavelet or scaling function, and *image box* in chapter 2 refers to any figure that contains projections and residuals of an original image after processing with wavelets. We feel these terms provide appropriately descriptive labels and use them without hesitation. The problems posed in this text are distributed throughout the reading rather than at the end of each chapter. This is important because completing the problems is vital to learning about wavelets. In fact, it is necessary to solve those problems marked with **bold** numbers to completely understand the text. Also, a computer algebra system is necessary to complete some problems, and hints and an appendix are provided for those readers who have access to *Maple*. Answers to selected problems are provided in an appendix. *Pixel Images* and *Maple* worksheets are available at our web site:

www.gvsu.edu/mathstat/wavelets.htm

Edward Aboufadel
Steven Schlicker

Allendale, Michigan
July, 1999

Acknowledgments

The authors would like to thank John Golden, Richard Gardner, Seth Falcon, and Phil Gustafson for their helpful comments on the manuscript. Special thanks goes to Matt Boelkins, who spent way too much time making comments on the drafts of this work.

We would also like to thank Christine Hughes, a mathematics student at GVSU who worked through an earlier draft of this book. Phil Pratt of GVSU gave us valuable advice on working with publishers.

Our heartfelt apologies go out to our spouses, Kathy and Deb, and to Steve's son Michael, for all the time that we spent working on this book which meant time away from them. Their patience has been appreciated.

If this book has a brother, it would be Ed's son, Zachary, who was born at the time we were finishing *Discovering Wavelets*.

E. F. A. and S. J. S.

Contents

1

Wavelets, Fingerprints, and Image Processing

1.1 PROBLEMS OF THE DIGITAL AGE

With the advent of the Digital Age, many opportunities have arisen for the collection, analysis, and dissemination of information. Dealing with such massive amounts of data presents difficulties. All of this digital information must be stored and be retrievable in an efficient manner. One approach to deal with these problems is to use *wavelets*. For example, the FBI fingerprint files contain over 25 million cards, each of which contains 10 rolled fingerprint impressions. Each card produces about 10 megabytes of data. To store all of these cards would require some 250 terabytes of space. Without some sort of compression of the data, the size of this database would make sorting, storing, and searching nearly impossible. To deal with this problem, the FBI has adopted standards for fingerprint digitization using a wavelet compression standard [3, 4, 7]. With wavelets, a compression ratio of about 20:1 is obtained.

Another common problem presented by electronic information is noise. Noise is extraneous information in a signal that can be introduced in the collection and transmission of data through a variety of means. Wavelets can be used to filter out noise via the computation of *averaging* and *detailing* coefficients. The detailing coefficients indicate the location of the details in the original data set. If some details are very small in relation to others, eliminating them may not substantially alter the original data set [11, 16]. Figure 1.1 illustrates a nuclear magnetic resonance (NMR) signal before and after denoising. Observe that the denoised data still demonstrates all of the important

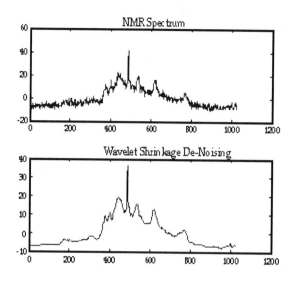

Fig. 1.1   "Before" and "after" illustrations of a nuclear magnetic resonance signal. (Images courtesy of David Donoho, Stanford University, NMR data courtesy Andrew Maudsley, VA Medical Center, San Francisco). Image copied from [16].

details. Similar ideas may be used to restore damaged video, photographs, or recordings.

Other applications of wavelets have emerged in seismology, astronomy, and radiology [19, 24]. It is with these varied applications in mind that we embark on our study of wavelets and their uses.

═══════════════ **Problems** ═══════════════

1. This book contains an extensive bibliography. Search the references, or find some of your own, to find two other applications of wavelets. Explain the problem to which wavelets are applied as best you can. Be sure to cite your source(s).

1.2   DIGITIZING FINGERPRINT IMAGES

As mentioned earlier, the United States Federal Bureau of Investigation (FBI) has collected the fingerprints of over 25 million people [3, 4, 7]. The first step in the wavelet compression process is to digitize each fingerprint image. There are two important ideas about digitization to understand here: intensity levels and resolution. In image processing, it is common to think of 256 different

Fig. 1.2   The gray scale, with intensity ranging from 255 down to 0.

Fig. 1.3   Pixels colored in scales of gray which are multiples of 10.

intensity levels, or scales, of gray, ranging from 0 (black) to 255 (white) as in figure 1.2. Each of these gray scales can be represented by an 8-bit binary number (e.g. 10101010 corresponds to the intensity level 170). A digital image can be created by taking a small grid of squares (called *pixels*) and coloring each pixel with some shade of gray as see in figure 1.3. The *resolution* of this grid is a measure of how many pixels are used per square inch. For fingerprints, the FBI uses a 500 dots per inch (or dpi) resolution, where the sides of each pixel measure 1/500th of an inch, so there are 250,000 pixels per square inch. (Typically, laser printers have a resolution of 300 or 600 dpi.) So, to digitize an image of one square inch at 500 dpi, a total of 8×250,000, or 2 million bits of storage space is needed.

Since 8 bits is the same as 1 byte, storing one square inch of image would require 250 kilobytes, roughly one-sixth of the memory on a typical floppy disk. Each fingerprint requires approximately 1.5 square inches, so the ten separate prints of a person would use nearly 4 megabytes, or about 3 floppy disks. However, the situation is really worse than this. Due to other prints taken by the FBI, such as simultaneous impressions of both hands, a fingerprint card for one person uses 10 megabytes of data. One of the reasons this is a problem is that, using a modem that transmits data at 56000 bits per second, it would take around a half an hour to send one card over a phone line.

The FBI could simplify the process by using a 1-bit scale (the pixel is either black or white), but they have found that "[8-bit] gray-scale images have a more 'natural' appearance to human viewers ... and allow a higher level of subjective discrimination by fingerprint examiners" [3]. For instance, the locations of sweat pores on the finger are legally admissible points of identification, and using an 8-bit scale permits the examiners to better see the pores [4]. Thus, the FBI faces the problem of too much information, and they need a way to process the data — in particular, to compress it — so that the storage and transmission of fingerprint images can be done quickly.

================================= **Problems** =================================

2. How much data (in megabytes) is generated by digitizing a 3 inch by 5 inch black-and-white photograph using 8-bit grayscale at 500 dpi?

3. To digitize a color picture requires 24 bits per pixel (8 bits for red, 8 for green, and 8 for blue), combined to make 16,777,216 different colors. How much data (in megabytes) is generated by digitizing a 3 inch by 5 inch color photograph using 24-bit colors at 500 dpi?

1.3 SIGNALS

In many situations we acquire data from measuring some phenomenon at various points. For example, the digitized information from fingerprints is such a collection of data, with each row of pixels considered as a separate group of data. Other examples of data collection include polling a small group of people on an issue to serve as a representation of the views of an entire group or measuring the results of an experiment at various times while it is running. We call this process of gathering data *sampling*. The information collected from sampling can be formed into strings of numbers called *signals*. We will represent signals as column vectors in $\mathbb{R}^n$. For example, $[12, 2, -5, 15]^T$ is such a signal (which, for instance, could result from measuring the temperature in Fahrenheit every 3 hours during a cold evening). (Note, the column vector $[12, 2, -5, 15]^T$ or $\mathbf{v}^T$ denotes the transpose of $\mathbf{v}$.) This signal is a vector in $\mathbb{R}^4$ and can be represented as a linear combination of basis vectors for $\mathbb{R}^4$. (Appendix A includes a review of the basics of linear algebra.)

================================= **Problems** =================================

4. Let

$$S_2 = \left\{ \begin{bmatrix} 1 \\ 0 \\ 0 \\ 0 \end{bmatrix}, \begin{bmatrix} 0 \\ 1 \\ 0 \\ 0 \end{bmatrix}, \begin{bmatrix} 0 \\ 0 \\ 1 \\ 0 \end{bmatrix}, \begin{bmatrix} 0 \\ 0 \\ 0 \\ 1 \end{bmatrix} \right\}$$

be the standard basis for $\mathbb{R}^4$. Express the signal $[12, 2, -5, 15]^T$ as a linear combination of the elements of S_2.

5. Let

$$
B_2 = \left\{ \begin{bmatrix} 1 \\ 1 \\ 1 \\ 1 \end{bmatrix}, \begin{bmatrix} 1 \\ 1 \\ -1 \\ -1 \end{bmatrix}, \begin{bmatrix} 1 \\ -1 \\ 0 \\ 0 \end{bmatrix}, \begin{bmatrix} 0 \\ 0 \\ 1 \\ -1 \end{bmatrix} \right\}.
$$

Show that B_2 is a basis for $\mathbb{R}^4$. Express the signal $[12, 2, -5, 15]^T$ as a linear combination of the basis elements from B_2.

Consider again the digitized data from a fingerprint. An image of one square inch using 500 dpi will yield 500 signals, where each signal is a vector in $\mathbb{R}^{500}$. Due to the 8-bit grayscale, these signals contain integers between 0 and 255.

What is it about wavelets that made them a natural choice to solve the problem of processing data from fingerprint images? Wavelets are useful when the signals being considered have parts where the data is relatively constant, while the changes between these constant parts are relatively large [16]. Fingerprints fit this description in that there are places where the intensity level is about constant, such as the white spaces between the ridges of the print or the ridges themselves, while the transition from a white space to a ridge involves a significant drop in intensity level (typically from 240 to 40). Later, we will see how the wavelet decomposition can filter data by averaging and detailing. Averaging is effective for areas of an image of relatively constant intensity, while detailing is effective in dealing with a sudden change of intensity.

1.4   THE HAAR WAVELET FAMILY

Wavelets are grouped into families, with names such as the Mexican Hat wavelets or the Shannon wavelets [21, 42]. All of these families have a number of common characteristics. The simplest one in which to see these characteristics clearly is the Haar family of wavelets. (The first mention of these wavelets appeared in an appendix to the thesis of A. Haar in 1909 [18], although the word *wavelet* was not coined until the 1980s.). While Haar wavelets are not really used in the applications discussed above, they will be used in the first two chapters of this text to demonstrate the fundamental ideas behind wavelets. Other wavelet families appear in subsequent chapters.

As will be the case with all families of wavelets described in this text, the Haar family is defined by two wavelets, a *father wavelet* and a *mother wavelet*. These wavelets are represented by ϕ and ψ, respectively. The father wavelet is usually referred to in the literature as the *scaling function*. The *Haar father wavelet* is defined by

$$
\phi(t) = \begin{cases} 1, & \text{if } 0 \le t \le 1 \\ 0, & \text{otherwise.} \end{cases}
$$

This father wavelet is also known as the *characteristic function* of the unit interval.

6. Use a Computer Algebra System (CAS) (for example, *Maple, Mathcad,* or *Mathematica*), to plot ϕ.

   ***Maple* Hint:** To define ϕ in *Maple* we can use the piecewise command

```
> phi := t -> piecewise(0<=t and t<1,1,t<0 or t>=1,0);
```

The *Haar mother wavelet* is defined by

$$\psi(t) = \begin{cases} 1, & \text{if } 0 \le t < \frac{1}{2} \\ -1, & \text{if } 0 \le \frac{1}{2} < 1 \\ 0, & \text{otherwise.} \end{cases}$$

Note that the father and mother wavelets are related in the following way:

$$\psi(t) = \phi(2t) - \phi(2t - 1). \tag{1.1}$$

As suggested by the terminology, if there is a father wavelet, a mother wavelet, and a family of wavelets, then there ought to be children. The following two wavelets are the *first generation of daughters*:

$$\psi_{1,0}(t) = \psi(2t) \quad \psi_{1,1}(t) = \psi(2t - 1).$$

(The sons will be introduced in chapter 2.) Although the daughters appear to be derived only from the mother, from (1.1) we can see that they can also be defined in terms of the father wavelet, namely by

$$\psi_{1,0}(t) = \phi(4t) - \phi(4t - 1) \quad \psi_{1,1}(t) = \phi(4t - 2) - \phi(4t - 3).$$

7. Use a CAS to define and plot ψ, $\psi_{1,0}$, and $\psi_{1,1}$.

The graphs generated in problems 6 and 7 give a sense of why the term *wavelet* is used, for the graphs look like square waves. (Wavelet literally means "small wave", and comes from the French term *ondelette*.) These Haar wavelets have one particularly desirable property: they are 0 everywhere except on a small interval. (Chapter 3 shows why this is important.) If a function is 0 everywhere outside of a closed, bounded interval, the function has *compact support* (closed and bounded intervals are said to be compact). If a function has compact support, the smallest closed interval on which the function has nonzero values is called the *support* of the function.

=========================== **Problems** ===========================

8. Are the Haar wavelets continuous? Are their first derivatives continuous? Explain.

What is the connection between the Haar wavelets and linear algebra? As a starting point, a one-to-one correspondence between the wavelets ϕ, ψ, $\psi_{1,0}$ and $\psi_{1,1}$ and vectors in $\mathbb{R}^4$ can be defined as follows [33]:

$$\phi \leftrightarrow \begin{bmatrix} 1 \\ 1 \\ 1 \\ 1 \end{bmatrix} \quad \psi \leftrightarrow \begin{bmatrix} 1 \\ 1 \\ -1 \\ -1 \end{bmatrix} \quad \psi_{1,0} \leftrightarrow \begin{bmatrix} 1 \\ -1 \\ 0 \\ 0 \end{bmatrix} \quad \psi_{1,1} \leftrightarrow \begin{bmatrix} 0 \\ 0 \\ 1 \\ -1 \end{bmatrix}.$$

The correspondence becomes clear by looking at the graphs of these wavelets. For example, the graph of $\psi_{1,1}$ is shown in figure 1.4. (Note that *Maple* connects the pieces of the function with vertical segments. Although these vertical segments are not part of the graph, they are included here to make the resulting graph look more like a wave.) The first entry (0) of the vector associated

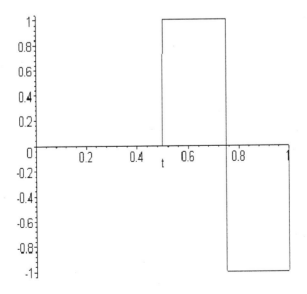

Fig. 1.4   The wavelet $\psi_{1,1}$.

to $\psi_{1,1}$ is the value of the wavelet on the interval $\left[0, \frac{1}{4}\right)$, the second entry (0) is the value on the wavelet on the interval $\left[\frac{1}{4}, \frac{1}{2}\right)$, the third entry (1) is the value of $\psi_{1,1}$ on $\left[\frac{1}{2}, \frac{3}{4}\right)$ and the final component (-1) is the value $\psi_{1,1}$ assumes on $\left[\frac{3}{4}, 1\right)$. There is a similar correspondence for each of the wavelets.

In fact, this correspondence can be extended to all functions which are "constant on quarters." To demonstrate this, let us define the set V_2 to consist of all functions which are piecewise constant on the intervals $\left[0, \frac{1}{4}\right)$, $\left[\frac{1}{4}, \frac{1}{2}\right)$, $\left[\frac{1}{3}, \frac{3}{4}\right)$, $\left[\frac{3}{4}, 1\right)$, and are zero outside of the interval $[0,1)$. Then we have the following correspondence between V_2 and $\mathbb{R}^4$:

$$
f(t) = \begin{cases} a, & \text{if } 0 \le t < \frac{1}{4} \\ b, & \text{if } \frac{1}{4} \le t < \frac{1}{2} \\ c, & \text{if } \frac{1}{2} \le t < \frac{3}{4} \\ d, & \text{if } \frac{3}{4} \le t < 1 \\ 0, & \text{otherwise} \end{cases} \quad \leftrightarrow \quad \begin{bmatrix} a \\ b \\ c \\ d \end{bmatrix}.
$$

This correspondence uniquely identifies a given element in V_2 with an element in $\mathbb{R}^4$. This means that the correspondence defines a function from V_2 to $\mathbb{R}^4$, that no two elements in V_2 are identified with the same vector in $\mathbb{R}^4$ (so the function is *one-to-one* or is an *injection*), and that each element in $\mathbb{R}^4$ corresponds to a function in V_2 (the function is *onto* or is a *surjection*).

Note that these vectors are *not* the wavelets, but rather that there is a correspondence between the wavelets and the vectors.

=================== **Problems** ===================

9. Sketch the graph of the function that corresponds to the following vector: $[2, -3, 5, 7]^T$.

In a similar fashion, let us define V_1 to be the set of functions which are piecewise constant on the intervals $\left[0, \frac{1}{2}\right)$ and $\left[\frac{1}{2}, 1\right)$, and are zero outside of the interval $[0, 1)$, and V_0 to be composed of functions which are constant on $[0,1)$ and zero outside that interval. Then there is a one-to-one correspondence between V_1 and $\mathbb{R}^2$, and between V_0 and $\mathbb{R}$, as well.

These ideas can be extended to other generations of wavelets and similar one-to-one correspondences. For example, the second generation of daughters is made up of the following four wavelets:

$$
\begin{array}{ll} \psi_{2,0}(t) = \psi(4t) & \psi_{2,1}(t) = \psi(4t - 1) \\ \psi_{2,2}(t) = \psi(4t - 2) & \psi_{2,3}(t) = \psi(4t - 3). \end{array}
$$

These four daughters are "constant on eighths". In general, the nth generation of daughters will have 2^n wavelets defined by

$$\psi_{n,k}(t) = \psi(2^n t - k), \quad 0 \le k \le 2^n - 1. \tag{1.2}$$

Observe that members of this generation will be constant on intervals of length $2^{-(n+1)}$.

================================ **Problems** ================================

10. Sketch the graphs of the second generation of wavelet daughters.

11. Use a CAS to plot some of the functions $\psi_{n,k}$.

> ***Maple* Hint:** To define the functions $\psi_{2,k}$, use a loop and define ψ as a function. (Refer to how ϕ was defined earlier in problem 6.) The following loop will then define each $\psi_{2,k}$ as an expression.

```
> n := 2;
> for k from 0 to 2^n-1 do
> psi.n.k := psi(2^n*t-k):
> od:
```

> Be careful about the difference between *functions* and *expressions* in *Maple*. In order to change the expression psi.n.k to the function psi.n.k, use the following command.

```
> psi.n.k := unapply(psi.n.k, t):
```

This leads naturally to more sets of functions and more correspondences. The set V_3 contains the functions which are "constant on eighths", in the sense that each function in V_3 is constant on $\left[0, \frac{1}{8}\right)$, $\left[\frac{1}{8}, \frac{1}{4}\right)$, and so on. There is a correspondence between V_3 and $\mathbb{R}^8$, where entries in a vector will be the values of a function on $\left[0, \frac{1}{8}\right)$, $\left[\frac{1}{8}, \frac{1}{4}\right)$, etc.

================================ **Problems** ================================

12. In general, how would V_n be defined? On what length intervals are the functions in V_n constant? For what value of m will V_n correspond to $\mathbb{R}^m$?

1.5 PROCESSING SIGNALS

A key idea in the study of wavelets is that functions that belong to V_2 can be written as linear combinations of the father and mother wavelets and the first generation of daughters. For example, consider the function f defined by

$$f(t) = \begin{cases} -5, & \text{if } 0 \le t < \frac{1}{4} \\ -1, & \text{if } \frac{1}{4} \le t < \frac{1}{2} \\ 1, & \text{if } \frac{1}{2} \le t < \frac{3}{4} \\ 11, & \text{if } \frac{3}{4} \le t < 1 \\ 0, & \text{otherwise} \end{cases} \leftrightarrow \begin{bmatrix} -5 \\ -1 \\ 1 \\ 11 \end{bmatrix}.$$

Are there unique coefficients x_1, x_2, x_3, x_4 so that

$$f(t) = x_1\phi(t) + x_2\psi(t) + x_3\psi_{1,0}(t) + x_4\psi_{1,1}(t)? \tag{1.3}$$

This question can be rephrased in terms of elementary linear algebra. Using our identification of V_2 with $\mathbb{R}^4$ we can write (1.3) in the form

$$\begin{bmatrix} -5 \\ -1 \\ 1 \\ 11 \end{bmatrix} = x_1\begin{bmatrix} 1 \\ 1 \\ 1 \\ 1 \end{bmatrix} + x_2\begin{bmatrix} 1 \\ 1 \\ -1 \\ -1 \end{bmatrix} + x_3\begin{bmatrix} 1 \\ -1 \\ 0 \\ 0 \end{bmatrix} + x_4\begin{bmatrix} 0 \\ 0 \\ 1 \\ -1 \end{bmatrix}.$$

Equivalently, if we define

$$A_2 = \begin{bmatrix} 1 & 1 & 1 & 0 \\ 1 & 1 & -1 & 0 \\ 1 & -1 & 0 & 1 \\ 1 & -1 & 0 & -1 \end{bmatrix} \quad \text{and} \quad \mathbf{b} = \begin{bmatrix} -5 \\ -1 \\ 1 \\ 11 \end{bmatrix},$$

then the question becomes: does $A_2\mathbf{x} = \mathbf{b}$ have a unique solution? Since the vectors corresponding to the wavelets ϕ, ψ, $\psi_{1,0}$, and $\psi_{1,1}$ are linearly independent (in fact, these vectors are identified with the basis B_2 for $\mathbb{R}^4$), the answer to both questions is "yes". The linear independence of these vectors makes A_2 an invertible matrix; in particular:

$$A_2^{-1} = \begin{bmatrix} 0.25 & 0.25 & 0.25 & 0.25 \\ 0.25 & 0.25 & -0.25 & -0.25 \\ 0.5 & -0.5 & 0 & 0 \\ 0 & 0 & 0.5 & -0.5 \end{bmatrix} \quad \text{and} \quad \mathbf{x} = A_2^{-1}\mathbf{b} = \begin{bmatrix} 1.5 \\ -4.5 \\ -2 \\ -5 \end{bmatrix}.$$

The vector $\mathbf{b}$, or $f(t)$, forms the signal, and the numbers 1.5, -4.5, -2, and -5, obtained from the solution of $A_2\mathbf{x} = \mathbf{b}$, are called *wavelet coefficients*. The act of solving $A_2\mathbf{x} = \mathbf{b}$ is called *decomposing a signal into wavelet coefficients*, and is a critical step in the processing of data with Haar wavelets. The reverse process is called *recomposing a signal from its wavelet coefficients*. In this case, vector $\mathbf{x}$ is known and used to find vector $\mathbf{b}$. For recomposing, simply perform the multiplication $A_2\mathbf{x}$. The matrix A_2 is called the *Haar wavelet matrix* for $n=2$.

_____ **Problems** _____

13. (a) Decompose the following signals into wavelet coefficients. Then
     recompose the signal from wavelet coefficients:

$$(i) \quad [3, 7, -4, -6]^T \quad (ii) \quad [14, 44, -25, -25]^T.$$

(b) Suppose x_1, x_2, x_3, and x_4 are the wavelet coefficients for the signal
$[3, 7, -4, -6]^T$ computed in (a). Plot the expression

$$x_1\phi + x_2\psi + x_3\psi_{1,0} + x_4\psi_{1,1}.$$

What do you get?

_____

Since B_2 is a basis for $\mathbb{R}^4$, a unique solution to (1.3) is obtained for every
f in V_2. This gives the following theorem.

Theorem. Every function in V_2 can be written uniquely as a linear combination of ϕ, ψ, $\psi_{1,0}$, and $\psi_{1,1}$.

In fact, V_2 is a vector space under pointwise addition and the standard
scalar multiplication of functions.

_____ **Problems** _____

14. What is the zero vector of V_2? What is the dimension of V_2? Explain
why it is correct to say that the family of wavelets

$$B_2 \ = \ \{\phi, \psi, \psi_{1,0}, \psi_{1,1}\}$$

is a basis of V_2. Note carefully that the same label is used for this basis
B_2 as was used for the basis of $\mathbb{R}^4$ defined in problem 5. This use of the
same label for two different bases may be confusing at first. However,
there is an identification between the elements of the two bases, so this
use of notation is reasonable. It should be clear from the context exactly
which basis we mean.

15. For each n, the set V_n may also be considered a vector space. What is
the dimension of V_3? Find a basis B_3 for V_3 using father, mother, and
daughter wavelets. (See the note about labeling in problem 14.) Create
the corresponding matrix A_3.

16. (a) Decompose the following signals obtained from functions in V_3 into
     wavelet coefficients:

     i. $[3, 7, -4, -6, 14, 44, -25, -25]^T$

      ii. $[250, 250, 240, 220, 15, 5, 5, 4]^T$

(b) Select one of the signals from (a). Let $x_1, x_2, \ldots, x_8$ be the wavelet coefficients for the signal computed in part (a). Plot
$$x_1\phi(t) + x_2\psi(t) + x_3\psi_{1,0}(t) + x_4\psi_{1,1}(t) +$$
$$x_5\psi_{2,0}(t) + x_6\psi_{2,1}(t) + x_7\psi_{2,2}(t) + x_8\psi_{2,3}(t).$$
What do you get?

(c) Recompose both signals in (a) from their wavelet coefficients.

17. What is the dimension of V_n? Find a basis B_n for V_n using father, mother, and daughter wavelets. (See the note about labeling in problem 14.) What is the size of the matrix A_n? For which m is the space V_n isomorphic to the vector space $\mathbb{R}^m$?

1.6   THRESHOLDING AND COMPRESSION OF DATA

How does the use of wavelets save space in storing and transmitting data? When sending data electronically, each bit costs time and money. As a result, current research is being conducted into how a given amount of information can be stored and transmitted in as few bits as possible. As will be shown, for certain signals many of the wavelet coefficients are close to or equal to zero. Through a method called *thresholding* [27], these coefficients may be modified so that the sequence of wavelet coefficients contains long strings of 0's. These long strings can be stored or sent electronically in much less space through a type of compression known as *entropy coding*.

In many situations, some of the information collected about an object tells very little about it. For example, consider the signal obtained from sampling the function f defined by $f(t) = \sin(20t)(\ln(t))^2$ at 32 evenly spaced points in [0,1]. (Note: that the sampling is on [0,1] is independent of the fact that the support of the Haar wavelets is also [0,1]. All we are doing is creating a signal with 32 entries. The interval itself is irrelevant.) A *Maple* plot of the graph of f is shown in figure 1.5. If f is sampled at 32 points and plotted by connecting the points in sequence, figure 1.6 results.

This function exhibits more variation in its values close to $t = 0$ than near $t = 1$. To have enough data to represent the essence of this function, it is more important to capture this variation near $t = 0$ than the lack of variation near $t = 1$. How can this be accomplished?

By separating the data into 4 vectors (signals), each with 8 entries, the data can be decomposed into wavelet coefficients using the matrix A_3 from problem 15. (Separating the signal like this is artificial, but it does allow processing with a smaller matrix.) The resulting vectors of wavelet coefficients are (entries rounded to the nearest thousandths)

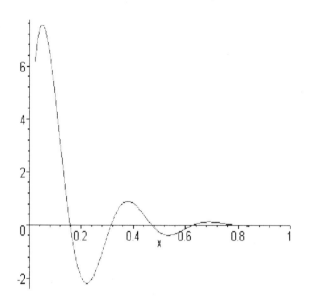

Fig. 1.5   The function $f(t) = \sin(20t)(\ln(t))^2$.

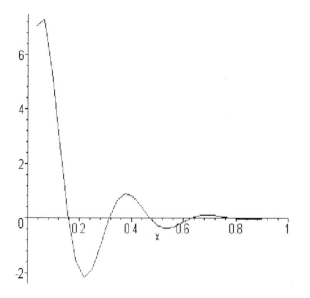

Fig. 1.6   Sampling $f(t)$ at 32 evenly spaced points.

$$[2.086, 3.478, 1.597, 0.620, -0.134, 1.380, 0.829, -0.168],$$

$$[0.186, -0.058, -0.642, 0.360, -0.470, -0.133, 0.177, 0.145],$$

$$[-0.062, -0.158, -0.126, 0.016, -0.026, -0.079, -0.018, 0.026], \text{ and}$$

$$[-0.008, -0.007, 0.007, -0.002, 0.014, -0.005, -0.003, 0].$$

Note that there are many wavelet coefficients that are quite small compared to others. In one sense, each coefficient indicates the extent of the "detail" about the whole picture that is contained by that particular piece of information. One can reasonably expect that little information will be lost if these small "detail" coefficients are ignored.

One way to ignore the small coefficients is through *hard thresholding* or *keep or kill*. In hard thresholding a tolerance, λ, is selected. Any wavelet coefficient whose absolute value falls below that tolerance is set to 0 with the aim to introduce lots of 0s (which can be compressed) without losing significant detail. There is no easy way to choose λ, though clearly with a larger threshold more coefficients are changed, introducing more error into the process.

To examine the effects of thresholding, let's process the coefficients in our example using a tolerance of 0.05. Each coefficient whose absolute value is less than 0.05 is replaced with 0. Recomposing a signal from these new coefficients gives us a string of 32 function values similar to the original signal. Figure 1.7 plots both the recomposed data and the original data. Note that there are two graphs plotted here. However, the graphs are so close it is nearly impossible to tell them apart.

Another way to ignore data is by *soft thresholding*. Again set a tolerance λ. If an entry is less than λ in absolute value, set that entry equal to 0. In addition, all other entries d are replaced with

$$\text{sign}(d)\big||d| - \lambda\big|.$$

We can think of soft thresholding as performing a translation of the signal toward zero by the amount λ.

A third method is called *quantile thresholding*. Select a percentage p of entries to be eliminated, and then set the smallest (in absolute value) p percent of the entries to zero.

Problems

18. Choose a function f different from the example used above. In selecting your function, be sure to consider the criteria discussed in this section. Sample, process, perform hard thresholding, and then reprocess the sampled data to compare with the original.

    (a) Explain why you chose the function you did.

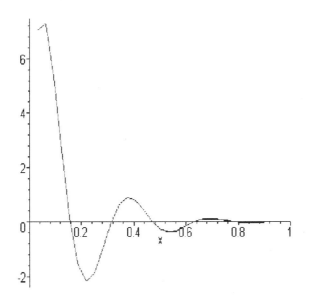

Fig. 1.7 Graphs of the original signal and the recomposed signal after hard thresholding with a tolerance of $\lambda = 0.05$.

(b) Sample your function at 32 evenly spaced points to obtain a signal, **s**, of length 32.

(c) Break **s** into signals of length 8 and compute the wavelet coefficients using the wavelet matrix A_3.

(d) Use hard thresholding to introduce strings of zeros into your processed signal. Reprocess the resulting signal. Experiment with different threshold levels and consider the number of zeros obtained in your processed signal after thresholding compared with the accuracy of the reconstructed signal.

(e) Decide on a "best" threshold level. Explain why you feel your threshold level is "best". Be explicit about how you processed your data. Include all details!

19. Using the same signal **s** as in problem 18, repeat the processing using A_2. (You will again need to break **s** into signals of smaller length. How long should these signals be if we use A_2 to process them?)

20. Using the same signal **s** as in problem 18, repeat the processing using A_4.

21. Compare the results of problems 18 to 20. Which processing seems to give the "best" results? Explain why.

22. Read some of the referenced material or search the Internet and find information on *lossless* and *lossy* compression. Explain the difference between the two.

The use of wavelets and thresholding serves to process the original signal, but, to this point, no actual compression of data has occurred. One method used to compress data is *Huffman entropy coding*. In this method, an integer sequence **q** is changed into a (hopefully) shorter sequence **e**, with the numbers in **e** being 8-bit integers (i.e., between 0 and 255). The entropy coding table (table 1.1) shows how the conversion is made. Numbers 1 through 100, 105, and 106 are used to code strings of zeroes, while 101 through 104 and 107 through 254 code the non-zero integers in **q**.

The codes 101 through 106 are used for larger numbers or longer zero sequences. The idea is to use two or three numbers for coding, the first being a flag to signal that a large number or long zero sequence is coming. For example, a string of 115 zeroes would be coded by "105 115", and the number 200 would be coded as "101 200". To code a number larger than 256 and less than 65,536, we first divide the number by 256. Since $65,536 = 256^2$, the quotient must be less than 256, and the remainder must be as well. In this way we can code these numbers with two 8-bit ones. For example, to code 10000, we first divide the number by 256 to get 39 with a remainder of 16. Then the coding would be "103 39 16". There is no provision for assigning a symbol to a number whose absolute value is greater than 65,535.

To illustrate, consider the signal $[210, 11, 0, 0, 0, -55, 4250, -5000]^T$. The first entry is coded as 101 210. The second is assigned the code 191 and the sixth is stored as 124. The third through fifth entries form a string of zeros that is coded as 3. Since

$$4250 = 16 \times 256 + 154 \quad \text{and} \quad -5000 = -(19 \times 256 + 136),$$

these entries are coded as "103 16 154" and "104 19 136" respectively. The new signal is $[101, 210, 191, 3, 124, 103, 16, 154, 104, 19, 136]^T$. Given that we began with a short signal, we should not expect to achieve any significant compression in this example. With long signals, however, like those arising from fingerprint images, the amount of space saved in this manner can be quite impressive.

Problems

23. Use entropy coding to compress the signal **q** into another signal **e**, where

$$\mathbf{q} = [-70, -40, 1, 0, 0, 0, 0, 0, 0, 34, 2001, 46, -34988, 0, 0, 0, 0, 0]^T.$$

Coding in e	Value in **q**
1	string of one zero
2	string of two zeroes
3	string of three zeroes
⋮	
100	string of 100 zeroes
101	number between 75 and 255, the exact number is next
102	number between -255 and -74, the absolute value of the exact number is next
103	number between 256 and 65535, the exact number is next (transmitted as two 8-bit integers)
104	number between -65535 and -256, the absolute value of the exact number (as two 8-bit integers) is next
105	string of zeroes between 101 and 255 zeroes, exact number is next
106	string of zeroes between 256 and 65535 zeroes, exact number is next (as two 8-bit integers)
107	-73
108	-72
109	-71
⋮	
179	-1
180	not used, use 1 instead
181	1
⋮	
253	73
254	74

Table 1.1   Entropy Coding Table

Entropy coding is used to code information more efficiently [36]. A coding table, such as the one in table 1.1, is designed so that the numbers that one expects would appear most often in **q** need the least amount of space in **e**. Table 1.1 was designed based on the assumption that **q** will mostly be made up of integers between -73 and 74, and strings of up to 100 zeroes. Although integers smaller than -73 and larger than 74 could appear, along with strings of more than 100 zeroes, they won't appear often, so the codes that need two numbers are reserved for those situations. This is similar to Morse code where the letters that appear most often in English words, such as E, T, I, A, and N are represented by only one or two symbols, while less used letters, such as X, Z, and Q, take four symbols.

1.7   THE FBI WAVELET/SCALAR QUANTIZATION STANDARD

The FBI has utilized wavelets to compress the digitized fingerprint images discussed in section 1.2. The method used by the FBI is referred to as the *wavelet/scalar quantization* (WSQ) standard, and is part of the Integrated Automated Fingerprint Identification System. According to the FBI:

> "The Federal Bureau of Investigation's (FBI's) Integrated Automated Fingerprint Identification System (IAFIS) is being developed to sustain the FBI's mission to provide identification services to the nation's law enforcement community and to organizations where criminal background histories are a critical factor in consideration for employment. The IAFIS will serve the FBI well into the twenty-first century and represents a quantum leap in communications, computing, and data storage and retrieval technologies." [3]

The basic steps used to compress fingerprints by the FBI are:

1. Digitize the source image into a signal **s** (a string of numbers).

2. Decompose the signal into a sequence of wavelet coefficients **w**.

3. Use thresholding to modify the wavelet coefficients from **w** to another sequence **w'**.

4. Use quantization to convert **w'** to a sequence **q**.

5. Apply entropy coding to compress **q** into sequence **e**.

The following example will demonstrate how this process works. The first step to compress fingerprints was discussed in section 1.2. Suppose we have a signal

$$s = [146, 134, 140, 140, 45, 41, 44, 2]^{T}.$$

One special characteristic of this sequence is that it contains subsequences with entries whose values are close to each other and it exhibits big jumps in values in other places, which makes it ideal for an application of wavelets.

Using the ideas from section 1.5, decompose the signal into wavelet coefficients using the Haar wavelets. Since the signal has length 8, use A_3 to solve the linear system $A\mathbf{x} = \mathbf{b}$ to obtain the solution

$$\mathbf{w} = [86.5, 53.5, 0, 10, 6, 0, 2, 21]^T.$$

The third step is to apply thresholding, as described in section 1.6, to modify $\mathbf{w}$. This is often thought of as filtering out some of the "noise" in the signal. Apply quantile thresholding with $p = 35$ (to insure 35% of the signal consists of 0s), to obtain

$$\mathbf{w'} = [86.5, 53.5, 0, 10, 6, 0, 0, 21]^T.$$

If hard thresholding with $\lambda = 11$ is applied instead, the result is

$$\mathbf{w'} = [86.5, 53.5, 0, 0, 0, 0, 0, 21]^T.$$

The fourth step in the process is *quantization*, a procedure which changes the sequence of floating-point numbers $\mathbf{w'}$ to a sequence of integers $\mathbf{q}$. For our work, each integer z must satisfy $-65535 \leq z \leq 65535$. The simplest form of quantization is to simply round to the nearest integer. Another possibility is to multiply each number in $\mathbf{w'}$ by some constant k and then apply rounding. More sophisticated quantization methods are also available, including the method used by the FBI [3]. Quantization is called *lossy* because it introduces error into the process, since the conversion from $\mathbf{w'}$ to $\mathbf{q}$ is not a one-to-one function. (Note that thresholding is also lossy.)

Returning to our example, let's use hard thresholding with $\lambda = 11$ followed by simple quantization with $k = 2$. The resulting sequence is

$$\mathbf{q} = [173, 107, 0, 0, 0, 0, 0, 42]^T.$$

Applying entropy coding to $\mathbf{q}$, we see that 173 is coded as "101 173", 107 becomes "101 107", and 42 changes to "222". The string of five zeroes is converted to "5". So, our final code is

$$\mathbf{e} = [101, 173, 101, 107, 5, 222]^T.$$

In this example $\mathbf{e}$ is 25% shorter than our original signal $\mathbf{s}$, yielding a compression ratio of 4:3. After compression, every number in $\mathbf{e}$ can be converted to an 8-bit binary number. This last string, $\mathbf{e}$, contains information that can be used to create a signal that is very close to our original signal $\mathbf{s}$ but can be stored in less space.

Although the process used by the FBI is more complicated than what was just described, the basic idea is the same. One significant difference is that the FBI processes a matrix of data from an image rather than from the one-dimensional signal $\mathbf{s}$. As a result, the FBI uses a wavelet family that has two father and two mother wavelets. These *symmetric biorthogonal wavelets,*

Fig. 1.8   An FBI-digitized left thumb fingerprint. The image on the left is the original; the one on the right is reconstructed from a 26:1 compression. (Courtesy of Chris Brislawn, Los Alamos National Laboratory).

which were developed by Cohen, Daubechies, and Feauveau [4], are beyond the scope of this book. (The interested reader is referred to [10], p. 259.)

For the process used by the FBI, each compressed image carries a tag with the compression rate for that image. The IAFIS is designed to allow for images to be compressed by WSQ at different compression rates. This is useful since an image of a "pinkie" will likely contain more white space (and consequently more zeros in the signal) than a thumb print image. This adaptive compression enables a typical little finger print to be preserved at a higher compression rate than a thumb print.

The effectiveness of the algorithm just discussed can be seen through an example. Figure 1.8 shows an FBI-digitized left thumb fingerprint and the result of a 26:1 compression of this thumbprint. The small details such as ridge endings and ridge textures are preserved. This image can be retrieved by anonymous FTP at

$$\texttt{ftp://ftp.c3.lanl.gov/pub/WSQ/print\_data/}$$

―――――――――――――――― **Problems** ――――――――――――――――

24. Consider the function f defined by $f(t) = e^{(-t^2)}$.

    (a) Sample f at 16 uniformly spaced points on [0,2]. Construct a signal from this data. (Note that we are sampling on [0,2] even though our wavelets are defined only on [0,1]. Explain why this does not pose a problem.)

    (b) Process the data with Haar wavelets to obtain wavelet coefficients. Choose your own matrix A_i when processing. You may want to experiment with different values of i.

    (c) Select a thresholding method and a threshold level and eliminate some of the processed data. Again, experiment with different threshold levels to achieve a reasonably accurate reconstruction of the original signal.

(d) Compress the data using steps 1 through 5 discussed in this section. Discuss the amount of compression achieved versus the quality of the graph reconstructed from the compressed data.

(e) Decide on a "best" level of compression and explain why you feel your chosen level is the best you can achieve.

2

Wavelets and Orthogonal Decompositions

2.1 A LEGO WORLD

In chapter 1, wavelets and their applications were introduced, as were the
vector spaces V_n of functions with compact support of $[0,1]$ which are constant
on intervals of length 2^{-n}. For each V_n, there is a basis, B_n, made up of
Haar wavelets. Now we are prepared for a more in-depth investigation of the
mathematics that drives wavelet theory.

================= **Problems** =================

1. A good way to "get your hands dirty" at this point is to get a box of
   Legos building blocks. (See [8] for an intriguing article about these toys.)
   Create, as accurately as you can, the graphs of the following functions
   on the interval $[0,1]$.

   (a) $f_1(t) = \sin(t)$
   (b) $f_2(t) = t^2$
   (c) $f_3(t) = e^t$
   (d) $f_4(t) = \sqrt{t}$

   This activity gives you a sense of what it means to approximate continuous functions with piecewise constant functions.

Problem 1 illustrates an important idea that will be encountered in this chapter. The spaces V_n contain only a few continuous functions. However, when we look at the functions in these spaces for larger and larger values of n, we see some functions that appear to be almost continuous. In fact, many of these functions begin to look like ones that are produced by a graphing calculator. By using functions in these spaces V_n, we will be able to approximate continuous functions much like our calculators do, but with even better accuracy. To make these ideas more concrete, let's examine the spaces V_n in more detail.

Before discussing the notion of approximating one function by another, the idea of "distance" between functions needs to be defined. This is done in V_n using an inner product. For any two functions f and g in V_n, the *inner product* of f and g is defined as

$$\langle f, g \rangle = \int_0^1 f(t)g(t)\ dt. \tag{2.1}$$

Using this inner product, length and distance can be defined. In particular, the *length* or *norm* of a function f in V_n is

$$\|f\| = \sqrt{\langle f, f \rangle} = \sqrt{\int_0^1 (f(t))^2\ dt}$$

and the *distance* between functions f and g is defined to be

$$\|f - g\| = \sqrt{\int_0^1 (f(t) - g(t))^2\ dt}.$$

(For more information about inner products, see appendix A.)

Problems

2. Compute the following inner products:

    (a) $\langle t, \phi(t) \rangle$

    (b) $\langle \psi_{1,0}(t), \psi_{1,1}(t) \rangle$

3. Compute the following norms:

    (a) $\|\psi_{1,0}(t)\|$

    (b) $\|\psi_{2,1}(t)\|$

4. Prove or disprove: ϕ and $\psi_{1,0}$ are orthogonal in V_2 using the inner product (2.1). (Recall that two vectors $\mathbf{u}$ and $\mathbf{v}$ in an inner product space are orthogonal if $\langle \mathbf{u}, \mathbf{v} \rangle = 0$.)

5. We saw in chapter 1 that $\{\phi, \psi, \psi_{1,0}, \psi_{1,1}\}$ is a basis for V_2. Prove that this basis is an orthogonal basis, but not an orthonormal basis. (Recall

that two vectors $\mathbf{u}$ and $\mathbf{v}$ in an inner product space are orthonormal if they are orthogonal and $\|\mathbf{u}\| = \|\mathbf{v}\| = 1$.)

Due to the inner product on V_n, this vector space is referred to as an *inner product space*. Another example of an inner product space that uses the inner product (2.1) is $L^2([0,1])$. This is the vector space of all functions

$$f : [0,1] \to \mathbb{R}$$

such that

$$\|f\| = \sqrt{\int_0^1 (f(t))^2 \, dt}$$

is finite.

_____ **Problems** _____

6. Prove that $t^{-1/4} \in L^2([0,1])$ and that $t^{-1/2} \notin L^2([0,1])$.

7. In chapter 1 we made the connection between a function f in V_n and a signal or vector $\mathbf{s}$ in $\mathbb{R}^m$, where $m = 2^n$. Let $\mathbf{s}$ be a vector in $\mathbb{R}^m$, and f the corresponding function in V_n. Show that

$$\|f(t)\| = \frac{\sqrt{\mathbf{s} \cdot \mathbf{s}}}{2^n}.$$

Conclude that every function f in V_n is also in $L^2([0,1])$.

8. Prove that, for every n, V_n is a subspace of $L^2([0,1])$.

Since $L^2([0,1])$, along with each V_n, is an inner product space, all of the facts that are known about inner product spaces can be applied. One particularly important result is the following:

The Orthogonal Decomposition Theorem. If W is a finite-dimensional subspace of an inner product space V, then any $\mathbf{v} \in V$ can be written uniquely as $\mathbf{v} = \mathbf{w} + \mathbf{w}_\perp$, where $\mathbf{w} \in W$ and $\mathbf{w}_\perp \in W^\perp$. (This theorem can be represented by $V = W \oplus W^\perp$.)

This theorem is used in the following way: suppose we have an inner product space V and a subspace W with an orthogonal basis $\{\mathbf{w}_1, \mathbf{w}_2, \ldots, \mathbf{w}_k\}$. Further, suppose that $\mathbf{v} \in V$. Then $\mathbf{w}$, as described in the Orthogonal Decomposition Theorem, is the vector

$$\mathbf{w} = \frac{\langle \mathbf{v}, \mathbf{w}_1 \rangle}{\langle \mathbf{w}_1, \mathbf{w}_1 \rangle} \mathbf{w}_1 + \frac{\langle \mathbf{v}, \mathbf{w}_2 \rangle}{\langle \mathbf{w}_2, \mathbf{w}_2 \rangle} \mathbf{w}_2 + \ldots + \frac{\langle \mathbf{v}, \mathbf{w}_k \rangle}{\langle \mathbf{w}_k, \mathbf{w}_k \rangle} \mathbf{w}_k = \sum_{i=1}^{k} \frac{\langle \mathbf{v}, \mathbf{w}_i \rangle}{\langle \mathbf{w}_i, \mathbf{w}_i \rangle} \mathbf{w}_i.$$

To find $\mathbf{w}_\perp$, simply compute $\mathbf{w}_\perp = \mathbf{w} - \mathbf{v}$. The vector $\mathbf{w}$ is called the *orthogonal projection* of $\mathbf{v}$ onto W. The vector $\mathbf{w}_\perp$ is called the *residual*. (Note: when this theorem is used in this text, sometimes V is $L^2([0,1])$ and sometimes is it is one of the wavelet spaces V_n.)

=================== **Problems** ===================

9. Determine the orthogonal projection of $h(t) = t$ onto V_2, using the basis B_2. Label your result $f(t)$, sketch a graph of $f(t)$, and sketch a graph of the residual, $h(t) - f(t)$.

==

As shall be shown, creating functions with Legos is analogous to dividing a function into two pieces — one (the orthogonal projection) that belongs to some wavelet space V_n and another (the residual) that belongs to $V_n^\perp$ — and then discarding the second piece.

2.2   THE WAVELET SONS

In chapter 1, vectors in $\mathbb{R}^m$ were identified with functions in V_n, where $n = 2^m$. This identification led to using the same notation for the corresponding bases

$$B_2 = \left\{ \begin{bmatrix} 1 \\ 1 \\ 1 \\ 1 \end{bmatrix}, \begin{bmatrix} 1 \\ 1 \\ -1 \\ -1 \end{bmatrix}, \begin{bmatrix} 1 \\ -1 \\ 0 \\ 0 \end{bmatrix}, \begin{bmatrix} 0 \\ 0 \\ 1 \\ -1 \end{bmatrix} \right\}$$

for $\mathbb{R}^4$ and

$$B_2 = \{\phi, \psi, \psi_{1,0}, \psi_{1,1}\}$$

for V_2. We will continue with this use of notation. It should be clear from the context, however, which basis is meant.

As was proved in problem 5 of this chapter, B_2 is an orthogonal basis for V_2. This basis consists of the father and mother wavelets, and the two daughters of the first generation. The next two sections will demonstrate how this basis arises from the Orthogonal Decomposition Theorem.

There is another basis of wavelets that is natural to consider. Recall that in chapter 1, S_2 was defined to be the standard basis for $\mathbb{R}^4$ (see problem 4, chapter 1). There must be a basis for V_2 consisting of functions corresponding to these standard unit vectors. This basis is called S_2 as well. It is not difficult to see that this S_2 contains the functions $\phi_{2,0}, \phi_{2,1}, \phi_{2,2}$, and $\phi_{2,3}$ defined by

$$\phi_{2,0}(t) = \phi(2^2 t) \qquad\qquad \phi_{2,2}(t) = \phi(2^2 t - 2)$$
$$\phi_{2,1}(t) = \phi(2^2 t - 1) \qquad \phi_{2,3}(t) = \phi(2^2 t - 3).$$

You will investigate these basis functions in the next problems.

_____ **Problems** _____

10. Verify the identifications between the two bases S_2 as defined in the preceding paragraph.

11. Use a CAS to plot the four functions in S_2. Explain how their graphs are related to the graph of ϕ.

12. Write ϕ as a linear combination of the elements of the basis S_2.

13. Graph the linear combination

$$2\phi_{2,0} - 3\phi_{2,1} + 17\phi_{2,2} + 30\phi_{2,3}.$$

Using the family analogy, these functions, $\phi_{2,0}, \phi_{2,1}, \phi_{2,2}, \phi_{2,3}$, are called the *wavelet sons*. In particular, these functions contained in S_2 are the *second generation of sons*. Now, there are two different bases for the wavelet space V_2: a basis of parents and daughters, and a basis of sons.

As in chapter 1, further generations can be defined by replacing the power of 2 with n. In general, for each positive integer n we define,

$$\phi_{n,k}(t) = \phi(2^n t - k)$$

for $0 \le k \le 2^n - 1$. Note the similarity to the definition of $\psi_{n,k}(t)$ in (1.2). For a given n, we will let S_n denote the set of 2^n functions $\{\phi_{n,k}\}_{k=0}^{2^n-1}$. As was seen for S_2, it is true that that S_n forms a basis for the inner product space V_n. The subsequent problems continue to explore the wavelet sons.

_____ **Problems** _____

14. Plot some of the functions $\phi_{n,k}$.

   ***Maple* Hint:** As in chapter 1, a loop may be used to define $\phi_{n,k}$. Here is an example for $n = 2$:

```
> n := 2;
> for k from 0 to 2^n-1 do
> phi.n.k := phi(2^n*t-k):
> od:
```

15. (a) Write ϕ as a linear combination of the elements of S_1.

   (b) Write each of the elements of S_1 as linear combinations of the elements of S_2.

   (c) Conclude from (a) and (b) that

$$\phi(t) = \phi(2t) + \phi(2t - 1).$$

(Note: This is a specific example of an important property of the father wavelet, namely that $\phi(t)$ satisfies a *dilation equation*, which is an equation of the form

$$\phi(t) = \sum_{k=-\infty}^{\infty} c_k \phi(2t - k).$$

In the case of the Haar wavelets, $c_0 = c_1 = 1$, and all other coefficients are 0. This equation will be investigated further when we study other wavelet families in chapter 3.)

16. For every function in S_3, determine the vector in $\mathbb{R}^8$ that corresponds to it.

17. Write $\phi_{2,0}$ as a linear combination of the elements of S_3.

18. Write a short essay on the following statement: "The functions in S_n form a basis for the vector space of all piecewise constant functions on $[0,1]$ that have possible breaks at the points

$$2^{-n}, 2 \cdot 2^{-n}, 3 \cdot 2^{-n}, \ldots, (2^n - 1) \cdot 2^{-n}."$$

19. Recall that we can view a string of data as a piecewise constant function obtained by partitioning $[0,1]$ into 2^n subintervals, where 2^n represents the number of sample points. Suppose we collect the following data: $[10, 13, 21, 55, 3, 12, 4, 18]$.

    (a) Explain how this data may be used to define a piecewise constant function f on $[0,1]$.

    (b) Express f as a linear combination of suitable functions $\phi_{n,k}$.

    (c) Plot this linear combination to verify that it corresponds to f.

20. Prove or disprove: for each n, S_n is an orthogonal basis for V_n.

21. Graph $\phi(2t - k)$ for three values of k other than 0 and 1. How are these graphs different from the graphs of the sons and daughters we have seen so far?

2.3   SIBLING RIVALRY: TWO BASES FOR V_n

An important observation to make at this point is, given the vector space V_n with natural basis S_n, for any integer $n \geq 0$, the collection of inner product spaces $\{V_n\}$ forms a nested sequence of subspaces. That is,

$$V_0 \subseteq V_1 \subseteq V_2 \subseteq \cdots . \tag{2.2}$$

This sequence is part of what is called a *multiresolution analysis*, which will be studied in greater detail in chapter 3.

======================= **Problems** =======================

22. Explain why $V_i \subseteq V_j$, if $i < j$.

23. Which of the problems in section 2.2 demonstate the structure described by (2.2)?

There is an important connection between orthogonality and the wavelet daughters. In the problems and discussion which follow, we will investigate how the space V_1 can be viewed in terms of V_0 and then see how the basis B_2, which was introduced somewhat arbitrarily in chapter 1, arises from the standard basis.

Since V_0 is a subspace of V_1, the Orthogonal Decomposition Theorem can be used to write each function $h \in V_1$ as $h = f + g$, where $f \in V_0$ and $g \in V_0^\perp$, that is,

$$V_1 = V_0 \oplus V_0^\perp. \tag{2.3}$$

As an example, the projection of $\phi_{1,0} \in S_1$ onto V_0 is

$$\frac{\langle \phi_{1,0}, \phi \rangle}{\langle \phi, \phi \rangle} \phi = \frac{1}{2}\phi,$$

as given by the Orthogonal Decomposition Theorem. The graph of this projection is shown in figure 2.1.

The residual in this case, namely the projection onto $V_0^\perp$, is the difference

$$\phi_{1,0} - \frac{1}{2}\phi = \frac{1}{2}\psi.$$

The graph of the residual is shown in figure 2.2.

======================= **Problems** =======================

24. (a) Determine the projection of $\phi_{1,1}$, the other element of S_1, onto V_0. Sketch the result.

(b) Calculate the residual for $\phi_{1,1}$ and sketch the result.

(c) Write each of the elements of S_1 (which are in V_1) as a sum of a function in V_0 and a function in $V_0^\perp$.

Fig. 2.1   Projection of $\phi_{1,0}$ onto V_0.

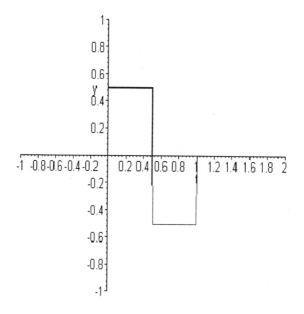

Fig. 2.2   Residual of $\phi_{1,0}$ onto V_0.

Note that V_1 has dimension 2, and that V_0 has dimension 1. By (2.3), $V_0^\perp$ must be one-dimensional. Further, the residuals of both $\phi_{1,0}$ and $\phi_{1,1}$ are scalar multiples of ψ. This shows that $\{\psi\}$ is a basis for $V_0^\perp$ and arises, via orthogonality, from the basis S_1. Note that ψ is, in fact, an element of V_1 and is orthogonal to everything in V_0.

As a result, we have another basis for V_1. Since $S_0 = \{\phi\}$ is a basis for V_0 and $C_0 = \{\psi\}$ is a basis for $V_0^\perp$, it follows from the Orthogonal Decomposition Theorem that

$$B_1 = S_0 \cup C_0 = \{\phi, \psi\}$$

is a basis for V_1.

Problems

25. Prove that $C_1 = \{\psi_{1,k} \mid \psi_{1,k}(t) = \psi(2t - k), k = 0, 1\}$ is a basis for $V_1^\perp$. Use the basis $B_1 = S_0 \cup C_0$ for V_1 to create a new basis for V_2. Where have we seen this basis before?

Since the set S_2 of wavelet sons in V_2 corresponds to the standard basis of $\mathbb{R}^4$, it might seem that these wavelets are the "best" with which to work. As we have seen, though, the basis S_2 just reads off the constant values of functions in V_2. This does not help us compress the information contained in these functions. However, the Orthogonal Decomposition Theorem shows that the familiar basis $B_2 = \{\phi, \psi, \psi_{1,0}, \psi_{1,1}\}$ arises quite naturally from S_2 (see problem 25).

This construction can be repeated for any value of n. Since V_n is a subspace of V_{n+1}, it follows that $V_{n+1} = V_n \oplus V_n^\perp$. As in problem 25, we can create a basis C_n of $V_n^\perp$ by

$$C_n = \{\psi_{n,k} \mid \psi_{n,k}(t) = \psi(2^n t - k), k = 0, 1, \ldots, 2^n - 1\}.$$

Proceeding inductively, and using the basis B_n of V_n, it follows that

$$B_{n+1} = B_n \cup C_n$$

is a basis for V_{n+1}. In general,

$$\begin{aligned}
V_n &= V_{n-1} \oplus V_{n-1}^\perp \\
&= (V_{n-2} \oplus V_{n-2}^\perp) \oplus V_{n-1}^\perp \\
&= \cdots \\
&= V_0 \oplus V_0^\perp \oplus V_1^\perp \oplus \cdots \oplus V_{n-1}^\perp.
\end{aligned} \tag{2.4}$$

This explains how the basis B_n is constructed from the "standard" basis S_n. As shown in chapter 1, the basis B_n is the one that supplies the wavelet coefficients.

=========================== **Problems** ===========================

26. List the functions in B_3 and B_4. (Note: Recall problems 15 and 17 in chapter 1.)

27. Write $\phi_{2,0}$ as a linear combination of the elements of B_3.

28. Explain the connection between problem 27 and (2.2).

29. As we have just seen, the basis B_2 for V_2 arises through the decomposition $V_2 = V_0 \oplus V_0^\perp \oplus V_1^\perp$. Compute the inner products of all pairs of basis elements in B_2. What kind of a basis is B_2? Why could this be useful? Explain.

Any function in V_n can now be expressed in two ways. If we favor the sons, then elements of V_n may be written as linear combinations of the functions in S_n. If, instead, we prefer to use the parents and daughters, then the result that

$$V_n = V_0 \oplus V_0^\perp \oplus V_1^\perp \oplus \cdots \oplus V_{n-1}^\perp$$

enables us to express functions in V_n as linear combinations of the functions in B_n. What has been gained from this sibling rivalry?

2.4 AVERAGING AND DIFFERENCING

Part of what results from the work in the previous section is a better understanding of the meaning of wavelet coefficients. In chapter 1, these coefficients arose from the solution to the linear equation $A\mathbf{x} = \mathbf{b}$. In this section, we will view these coefficients in terms of averages and differences, which will set the stage for a later discussion of filters. The following problems serve to introduce this approach.

=========================== **Problems** ===========================

30. The signal $[50, 16, 14, 28]^T$ represents a piecewise constant function in V_2 that can be written as

$$v = 50\phi_{2,0} + 16\phi_{2,1} + 14\phi_{2,2} + 28\phi_{2,3}.$$

Define another element of V_2 by

$$u = 33\phi_{1,0} + 21\phi_{1,1} + 17\psi_{1,0} - 7\psi_{1,1}.$$

Use each of the approaches below to show that $v = u$.

Method 1: Recall that $V_2 = V_1 \oplus V_1^{\perp}$ and that $\{\phi_{1,0}, \phi_{1,1}\}$ is a basis for V_1. In addition, remember that in problem 25, we saw that $\{\psi_{1,0}, \psi_{1,1}\}$ is a basis for $V_1^{\perp}$.

(a) Use the Orthogonal Decomposition Theorem to find the projection of v onto V_1.

(b) Hence, find vectors $v_1 \in V_1$ and $v_{1_{\perp}} \in V_1^{\perp}$ so that $v = v_1 + v_{1_{\perp}}$.

(c) Note the coefficients of v_1 and $v_{1_{\perp}}$ in terms of the bases for V_1 and $V_1^{\perp}$ that were given.

Method 2: View all of the vectors in this problem as elements of $\mathbb{R}^4$. Make sure you understand why

$$S_2 = \{\phi_{2,0}, \phi_{2,1}, \phi_{2,2}, \phi_{2,3}\}$$

and

$$D = \{\phi_{1,0}, \phi_{1,1}, \psi_{1,0}, \psi_{1,1}\}$$

each correspond to a basis for $\mathbb{R}^4$. With respect to S_2, the coefficients of v are $[50, 16, 14, 28]$. Use a change-of-basis matrix to find coefficients for v in terms of the basis D. What vector results?

31. Comment on the connection between the two methods in problem 30.

Note that something interesting happens in problem 30. The first coefficient (33) in u is the average of the first two coefficients (50 and 16) in v. The second coefficient (21) in u is the average of the second pair of coefficients (14 and 28) in v. The third coefficient (17) in u is the difference between 50 and 33 and the fourth coefficient (-7) in u is the difference between 14 and 21.

Problems

---

32. Using either method from the previous problem, express

$$33\phi_{1,0} + 21\phi_{1,1}$$

in V_1 as a linear combination of $\phi = \phi_{0,0}$ and $\psi = \psi_{0,0}$ in $V_0 \oplus V_0^{\perp}$. Does the same averaging and differencing pattern hold?

33. Explain why v from problem 30 may also be written as

$$27\phi + 6\psi + 17\psi_{1,0} - 7\psi_{1,1}$$

in $V_0 \oplus V_0^{\perp} \oplus V_1^{\perp}$.

Note that in the final decomposition, $27\phi + 6\psi + 17\psi_{1,0} - 7\psi_{1,1}$, the first coefficient is the overall average of the coefficients of the original linear combination $50\phi_{2,0} + 16\phi_{2,1} + 14\phi_{2,2} + 28\phi_{2,3}$, and the other coefficients are determined by computing the differences between successive averages and certain coefficients (and perhaps by dividing by 2). The process of finding these differences is also known as *detailing*.

The problems above demonstrate a general principle of wavelets. A member v of V_{n+1} can be written as a linear combination of the vectors in the standard basis. The projection of v onto V_n has new coefficients which are averages of original coefficients, while the coefficients of the residual of the projection are differences of the original coefficients. Repeating this process yields averages of the averages, and differences of the averages, and ultimately, the *wavelet coefficients* are created from this combination of averaging and differencing [26].

The coefficients in $27\phi + 6\psi + 17\psi_{1,0} - 7\psi_{1,1}$ are the wavelet coefficients of the original string because they are the coefficients of v with respect to the basis B_2. This time, though, they arise from various applications of averaging and differencing, rather than as entries in a column vector. This has important implications when considering an application such as image processing. If a certain area of a picture has constant intensity, computing differences (or differences of averages) will result in lots of zeros. As seen in chapter 1, this leads to effective data compression. Further, by thinking of wavelet coefficients in this way, we can develop other ways to study and compute them. Filters are one such approach which will be discussed in chapter 3.

Problems

34. Find a matrix M_4 so that $M_4 \begin{bmatrix} 50 \\ 16 \\ 14 \\ 28 \end{bmatrix} = \begin{bmatrix} 27 \\ 6 \\ 17 \\ -7 \end{bmatrix}$ by thinking of the process of averaging and differencing. The following steps will guide you.

   (a) The first transformation converted v in problem 30 to u. If we represent a generic vector v by $v = v_1\phi_{2,0} + v_2\phi_{2,1} + v_3\phi_{2,2} + v_4\phi_{2,3}$, then what would be the formula for u in terms of the v_i?

   (b) What would be the formula for the final decomposition (from problem 33) in terms of the v_i?

   (c) Use these results to create M_4. How is M_4 related to A_2 from chapter 1?

35. Let $v \in V_2$ be given by

$$v = 10\phi_{2,0} - 12\phi_{2,1} + 7\phi_{2,2} + 19\phi_{2,3}.$$

   Find wavelet coefficients so that

$$v = x_1\phi + x_2\psi + x_3\psi_{1,0} + x_4\psi_{1,1}$$

is in

$$V_0 \oplus V_0^{\perp} \oplus V_1^{\perp}.$$

For each of the four coefficients, explain how the averaging or differencing of the four numbers $10, -12, 7, 19$ leads to the coefficent.

36. Write a paragraph explaining how M_4 can be understood as a change-of-basis matrix.

37. Data is often stored in groups of 8. As in the previous problems, find a matrix M_8 that will perform the operations of averaging and differencing on an 8 by 1 column matrix. How is M_8 related to A_3 from chapter 1?

38. Use M_8 to find the wavelet coefficients for the signal

$$[80, 48, 4, 36, 28, 64, 6, 50]^T.$$

2.5 PROJECTING FUNCTIONS ONTO WAVELET SPACES

In this section, the ideas of Legos, orthogonal projections, and averaging and differencing are combined to create some interesting graphs. A good way to see these concepts together is through the following problem: given that the function h defined by $h(t) = 8t$ is in $L^2([0,1])$, and V_2 is a subspace of $L^2([0,1])$, determine the orthogonal projection of h onto V_2 using the basis B_2. How are the wavelet coefficients computed in this problem related to h?

The Orthogonal Decomposition Theorem shows that the projection of the function $h(t) = 8t$ onto V_2 is found by

$$\frac{\langle h, \phi \rangle}{\langle \phi, \phi \rangle}\phi + \frac{\langle h, \psi \rangle}{\langle \psi, \psi \rangle}\psi + \frac{\langle h, \psi_{1,0} \rangle}{\langle \psi_{1,0}, \psi_{1,0} \rangle}\psi_{1,0} + \frac{\langle h, \psi_{1,1} \rangle}{\langle \psi_{1,1}, \psi_{1,1} \rangle}\psi_{1,1} = 4\phi - 2\psi - \psi_{1,0} - \psi_{1,1}.$$

The graph of h and its projection onto V_2 are shown in figure 2.3.

An important point to recognize here is that the projection of h can be viewed as the signal obtained from sampling h at the midpoints of the quarter intervals (as discussed in section 1.6). The projection of h is identified with the signal

$$\left[h\left(\frac{1}{8}\right), h\left(\frac{3}{8}\right), h\left(\frac{5}{8}\right), h\left(\frac{7}{8}\right) \right] = [1, 3, 5, 7].$$

The wavelet coefficients of this projection, namely 4, -2, -1, and -1, arise from processing this signal as discussed in chapter 1. Recall the process: average the components of the signal in pairs, compute differences, then repeat. From this perspective, we can see that projecting onto the spaces V_n brings us back to the averaging and differencing we saw earlier.

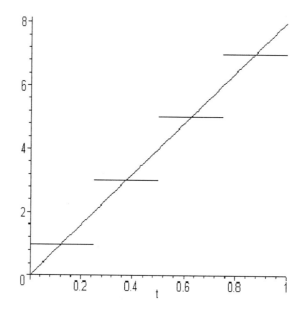

Fig. 2.3   Graphs of $8t$ and the projection of $8t$ onto V_2.

_____ **Problems** _____

39. Determine and graph the residual $g(t) = h(t) - f(t)$ for $h(t) = 8t$.

40. Develop formulas for the wavelet coefficients of the projection of $h(t) = 8t$ onto V_2 in terms of the average or difference of values of h.

From the previous example, we see that, while h lives in "the world of $L^2([0,1])$," its orthogonal projection f lives in "the Lego world," and that wavelet coefficients arise when the orthogonal projection is computed from averaging and differencing certain values of h. In the following exercises, we explore cases where h is nonlinear.

_____ **Problems** _____

41. Using the wavelets bases B_2, B_3, and B_4, determine and graph the projections of the following functions on V_2, V_3, and V_4.

   (a) $\sin(t)$
   (b) t^2
   (c) e^t
   (d) $\sqrt{t}$

(Note: you may wish to numerically approximate some of your integrals to save time.) What happens to the projections as you move from V_2 to V_3, and then to V_4?

42. Compare your results from 41(a) with sampling the sine function. At what points is this function being "sampled"? Develop formulas for the wavelet coefficients that you determined in terms of the average or difference of certain values of $\sin(t)$.

43. Write a short essay comparing your work in this section to one other application of orthogonal projections (e.g. Fourier series, regression lines).

2.6 FUNCTION PROCESSING AND IMAGE BOXES

In many books and articles about wavelets, pictures such as those in figure 2.4 are displayed. In this section, we will learn how these pictures are created.

We begin by returning to the example from chapter 1 of processing data sampled from a function. Recall that the function $f(x) = \sin(20x)(\ln x)^2$ was sampled at 32 evenly spaced points, generating the following data (rounded to the nearest thousandths):

$$[7.028, 7.300, 5.346, 2.588, 0.057, -1.602, -2.180, -1.843, -0.984,$$
$$-0.045, 0.636, 0.902, 0.782, 0.427, 0.029, -0.261, -0.373, -0.320,$$
$$-0.173, -0.015, 0.094, 0.130, 0.106, 0.054, 0.005, -0.022, -0.027,$$
$$-0.017, -0.006, 0, 0, 0].$$

This graph of the data is shown in figure 2.5.

Fig. 2.4   An image of a house, and an image box. (Image courtesy of Summus, Ltd.)

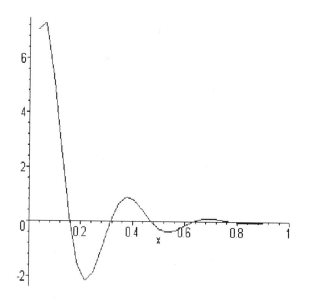

Fig. 2.5 A plot of the data.

In this section, the data is treated as one long string which connects the data points with line segments. To process this signal, we proceed in a fashion similar to the problems in section 2.4. The first step is to create averages (the coefficients of the projection of the signal onto V_4), and differences (the coefficients of the corresponding residual). This process yields 16 averages and 16 differences, which form one new signal:

$$[7.161, 3.967, -0.772, -2.011, -0.515, 0.769, 0.604, -0.116, -0.347,$$
$$-0.094, 0.112, 0.080, -0.009, -0.022, -0.003, 0,$$

$$-0.134, 1.380, 0.830, -0.168, -0.470, -0.133, 0.177, 0.145, -0.026,$$
$$-0.079, -0.018, 0.026, 0.0136, -0.005, -0.003, 0].$$

Graphing this data obtained from the first round of processing yields the graph in figure 2.6. We will call a figure that contains projections and residuals of an original image an *image box*. Examine this image box carefully. Note that the left half of the figure, created from the projection onto V_4, is a rough copy of the original data in half scale (see figure 2.5). The right half (the residual) shows how far the processed data is from the original.

The next step is to compute averages and differences on only the first half of the new data, leaving the residual alone. This projects the new signal (the first half of the data obtained after the first round of averaging and differencing) onto $V_3 \oplus V_3^{\perp}$. The resulting signal is:

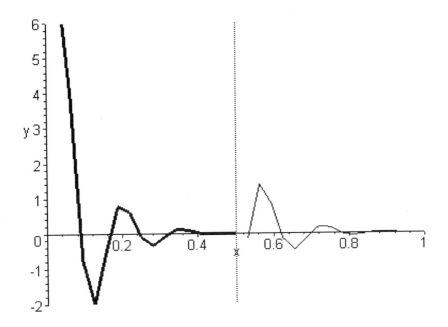

Fig. 2.6 An image box, showing the projections of the signal onto V_4 and $V_4^{\perp}$.

$[5.564, -1.392, 0.127, 0.244, -0.220, 0.096, -0.015, -0.002,$

$1.597, 0.620, -0.642, 0.360, -0.126, 0.016, 0.007, -0.002,$

$-0.134, 1.380, 0.830, -0.168, -0.470, -0.133, 0.177, 0.145, -0.026,$
$-0.079, -0.018, 0.026, 0.0136, -0.005, -0.003, 0].$

Plotting this data yields the image box in figure 2.7. Observe that the first quarter of this second-stage data contains a copy of the original data on a reduced scale. The projection onto V_3 appears to be the original graph compressed horizontally by a factor of 4. The second quarter (the residual from the second round of processing, which is in $V_3^{\perp}$) keeps track of how far the projection onto V_3 is from the processed data from the first stage. The last half of the data (the projection onto $V_4^{\perp}$) retains information about how far the processed figure from the first stage is from the original. We can continue this processing on finer and finer scales until we run out of things to process, yielding a final image box which is a graph of the signal containing the wavelet coefficients.

This example demonstrates how, as a signal is processed, imperfect copies of the original signal are made, along with other numbers which keep track of how far the copies are from the original signal.

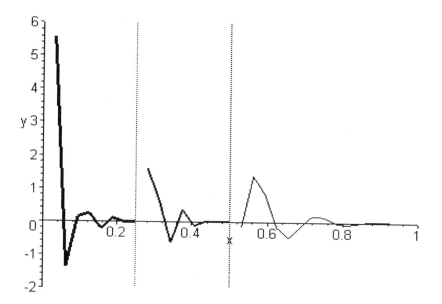

Fig. 2.7 An image box, containing the projections onto V_3, $V_3^\perp$, and $V_4^\perp$.

Problems

44. Select a function in $L^2[0, 1]$, built from combining at least two transcendental functions, and create an image box similar to figure 2.7.

45. Acquire a real signal whose length is a power of 2. (Some possiblities: use a Calculator-Based Laboratory to measure motion or temperature; stock market data; population trends.) Create an image box similar to figure 2.7 for that signal.

The image box in figure 2.4 is created in a similar way; the difference being that images are two-dimensional, whereas signals are one-dimensional. Because of this difference, we must employ a "trick" (Daubechies' word, see [10]) in order to use the Haar wavelets to process the image: we must think of the image as a matrix of numbers J, rather than a one-dimensional signal **s**. Suppose that J is a 32-by-32 matrix.

Begin by treating each of the rows as a separate signal of length 32 (in other words, as an element of V_5), and process these rows as before.[1] This

---

[1]In part 4 of the image compression project in section 4.2, the directions state to process the columns of J first. Processing the columns is easier to do than processing the rows, from the point of view of applying matrix algebra to the study of wavelets. However, when image boxes are presented in the literature, they are generated from processing the rows first.

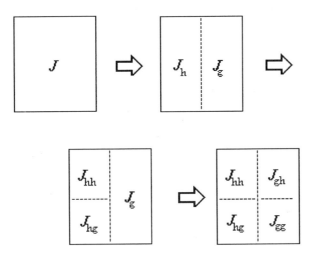

Fig. 2.8 Creating an image box.

results in two new 32-by-16 matrices J_h and J_g, where each row of J_h is from V_4 and each row of J_g is from $V_4^\perp$. (See figure 2.8. The use of the subscripts h and g will be clarified in chapter 3.)

Now focus on the columns of J_h, each of which can be thought of as a signal of length 32 (i.e., in V_5). Process the columns. This gives two new 16-by-16 matrices J_{hh} and J_{hg}, where each column of J_{hh} is from V_4 and each column of J_{hg} is from $V_4^\perp$. In the same manner, process the columns of J_g to create J_{gh} and J_{gg}.

What has been done is to decompose the original image, the 32-by-32 matrix, into four smaller images, each which is 16-by-16. So, the four images in figure 2.4 are a decomposition of the original image in figure 2.4. This process, which is often referred to as a *two-dimensional wavelet transform*, may then be repeated with the smaller matrix (image) J_{hh}.

This is only one possible approach to processing two-dimensional images. The FBI uses another method of applying a more elaborate family of wavelets which has two father wavelets and two mother wavelets [3, 10].

2.7 A SUMMARY OF TWO APPROACHES TO WAVELETS

To conclude this chapter, let's compare the methods thus far developed for finding wavelet coefficients using Haar wavelets from chapters 1 and 2. Consider again the example from chapter 1 where f is the function defined by

$$f(t) = \begin{cases} -5, & \text{if } 0 \le t < \frac{1}{4} \\ -1, & \text{if } \frac{1}{4} \le t < \frac{1}{2} \\ 1, & \text{if } \frac{1}{2} \le t < \frac{3}{4} \\ 11, & \text{if } \frac{3}{4} \le t < 1 \\ 0, & \text{otherwise.} \end{cases}$$

We found unique coefficients (the wavelet coefficients) so that

$$f(t) = x_1 \phi(t) + x_2 \psi(t) + x_3 \psi_{1,0}(t) + x_4 \psi_{1,1}(t)$$

by using matrix algebra. We identified V_2 with $\mathbb{R}^4$, defined the matrix A_2, and introduced vectors $\mathbf{x}$ and $\mathbf{b}$, to write the previous equation in the form $A_2 \mathbf{x} = \mathbf{b}$. Since A_2 is an invertible matrix, unique wavelet coefficients were obtained by multiplying $\mathbf{b}$ by A_2^{-1}. In particular, we get

$$\mathbf{x} = \begin{bmatrix} 1.5 \\ -4.5 \\ -2 \\ -5 \end{bmatrix}.$$

In chapter 2, we used the Orthogonal Decomposition Theorem to project f onto the space V_2 generated by the functions $\phi, \psi, \psi_{1,0}$, and $\psi_{1,1}$. Although the development was in terms of subspaces of $L^2([0,1])$, we were ultimately required, in the case where $f \in V_2$, to find a matrix M_4 that performs the averaging and differencing that generates the wavelet coefficients. This matrix M_4 turned out to be

$$\begin{bmatrix} 0.25 & 0.25 & 0.25 & 0.25 \\ 0.25 & 0.25 & -0.25 & -0.25 \\ 0.5 & -0.5 & 0 & 0 \\ 0 & 0 & 0.5 & -0.5 \end{bmatrix}.$$

We found the wavelet coefficients $\mathbf{x}$, via $\mathbf{x} = M_4 \mathbf{b}$. The process can be reversed through $\mathbf{b} = A_2 \mathbf{x}$, since the matrix M_4 is the inverse of A_2. Even in the case where $f \notin V_2$, we saw how the wavelet coefficients are based on averaging and differencing.

This gives us two perspectives on the same problem. In the first chapter, the focus is on solving a linear system to get coefficients in terms of a given basis. In the second chapter, the focus is to use orthogonality to create that basis and to investigate the projections of functions onto a sequence of inner product spaces. For the Haar wavelets, the approach we take depends on the function f, and whether we decide to work in the space $L^2([0,1])$ or the space $\mathbb{R}^n$ for a suitable n. It is not the case, however, that we can exploit this duality when working with other wavelets. Almost all other work with wavelets takes place in the space $L^2(\mathbb{R})$.

3

Multiresolutions, Cascades, and Filters

3.1 EXTENDING THE HAAR WAVELETS TO THE REAL LINE

In our work so far, we have considered functions only on the interval $[0,1]$. This is not the most convenient setting for signals. In this chapter, we broaden our perspective to develop wavelets as functions defined on all of $\mathbb{R}$. This will give us the freedom to manipulate signals that arise from functions that are defined anywhere on $\mathbb{R}$.

To work in this broader setting, we will need to extend the definitions of the Haar spaces V_n introduced earlier. For example, the space V_0 will become the set of piecewise constant functions with compact support that have possible breaks at integer values. A basis for V_0 will then be $\{\phi(t-k) : k \in \mathbb{Z}\}$.

Of course, it will not be enough to extend the definition of only V_0. We will redefine each of the spaces V_n to contain piecewise constant functions, defined on all of $\mathbb{R}$, *with compact support*. The space V_1 will consist of functions having compact support with possible breaks at rational points with denominators of 2 (i.e., the integers and the points midway between the integers). Similarly, V_2 will contain all functions with possible breaks at rational points with denominators of $4 = 2^2$ (i.e., the integers, the points midway between the integers, and the points one quarter of the way between the integers). In general, functions in V_n are piecewise constant having compact support with possible breaks at rational points of the form $\frac{m}{2^n}$, for any integer m.

From this point of view we can also allow n to be negative. In that case, V_n will contain piecewise constant functions having compact support with possible jumps at points of the form $m \times 2^{-n}$. Note that for negative n, the breaks in these functions will be farther apart rather than closer together.

In chapter 2 we defined bases for each of the spaces V_n. With those bases in mind, it is not hard to find spanning sets for the extended V_n. The old V_2 on $[0,1]$ had $\{\phi(2t), \phi(2t - 1)\}$ as a basis. This basis extends naturally to give us a spanning set $\{\phi(2t - k) : k \in \mathbb{Z}\}$ for the new V_2. In the same way, the set $\{\phi(2^n t - k) : k \in \mathbb{Z}\}$ is a spanning set for the new V_n. An important observation to make is that each function in our spanning set is simply a shifted version of the wavelets defined in chapter 2.

From this point on in the text, unless otherwise specified, any reference to the space V_n will refer to this new, extended version of V_n. Further, as in chapter 2, we will use the notation $\phi_{n,k}(t) = \phi(2^n t - k)$.

As we modify the spaces V_n, we will also need to alter the inner product. Recall that the inner product we have been using is

$$\langle f, g \rangle = \int_0^1 f(t)g(t)dt.$$

Since the wavelets in chapter 2 had values of 0 outside of the interval $[0,1]$, the restriction of the inner product to the interval $[0,1]$ is unnecessary. In fact, for any functions f, g in the old space V_n,

$$\langle f, g \rangle = \int_0^1 f(t)g(t)dt = \int_{-\infty}^{\infty} f(t)g(t)dt.$$

In this new perspective, where we consider functions defined on all of $\mathbb{R}$, we require a corresponding inner product, which is

$$\langle f, g \rangle = \int_{-\infty}^{\infty} f(t)g(t)dt. \tag{3.1}$$

When using this new inner product, it will be necessary to restrict ourselves to functions with finite norms. The set $L^2(\mathbb{R})$ is the collection of functions $f : \mathbb{R} \to \mathbb{R}$ such that

$$\|f\| = \langle f, f \rangle^{\frac{1}{2}} = \left(\int_{-\infty}^{\infty} f(t)^2 dt \right)^{\frac{1}{2}} < \infty.$$

This is also why we insisted that the spaces V_n contain only functions with compact support. Without that condition, the new V_n would not be a subspace of $L^2(\mathbb{R})$.

Observe that all of the old functions $\phi_{n,k}$ and $\psi_{n,k}$ from chapter 2, as well as any finite signal when viewed as a piecewise constant function, are elements of $L^2(\mathbb{R})$. This suggests that $L^2(\mathbb{R})$ is a more general setting in which to study wavelets.

_____ **Problems** _____

1. Find two continuous functions that belong to $L^2(\mathbb{R})$ and two that do not. What must be true about $\lim_{t \to \infty} f(t)$ and $\lim_{t \to -\infty} f(t)$ if f is a continuous function in $L^2(\mathbb{R})$? Explain.

2. (a) Provide an example to show that V_2 is not a subset of $L^2(\mathbb{R})$ if we don't include the condition that the functions in V_2 have compact support.

   (b) Show that, for any $n > 2$, V_n is not a subset of $L^2(\mathbb{R})$ if we don't include the condition that the functions in V_n have compact support.

3. A function in $L^2(\mathbb{R})$ is said to be *normalized* if the norm of the function is 1.

   (a) Find two functions in $L^2(\mathbb{R})$ that are normalized.

   (b) Let f be the function defined by $f(t) = \begin{cases} 1, & \text{if } 0 \le t \le 1 \\ -1, & \text{if } -1 \le t < 0 \\ 0, & \text{otherwise.} \end{cases}$

   Show that f is in $L^2(\mathbb{R})$ and calculate the norm of f. Show that $\frac{f}{\|f\|}$ is a normalized function in $L^2(\mathbb{R})$.

3.2   OTHER ELEMENTARY WAVELET FAMILIES

The Haar wavelets satisfy certain properties that are particularly important, as will be shown later. Each function $\phi(2^n t - k)$, $n, k \in \mathbb{Z}$ has compact support and is discontinuous on the real line. Further, $\langle \phi(2^n t), \phi(2^n t - k) \rangle = 0$ for every n and every nonzero k, and $\langle \phi(t), \phi(t) \rangle = 1$. These properties should be verified by the reader. The latter property shows that the norm of ϕ is 1 and hence ϕ is normalized.

In chapter 2, it was shown that all of the other wavelets could be obtained by combining scalings and translations of the father wavelet ϕ. For this reason, ϕ is referred to as the *Haar scaling function*.

_____ **Problems** _____

4. Use the definition of the inner product to prove that $\|\phi(mt - k)\| = \sqrt{\frac{1}{m}}$ for any $m = 2^n$, $n \in \mathbb{Z}$, and any integer k.

While the Haar wavelets are the simplest wavelets to understand and work with, there are many other wavelet families. As was the case with the Haar wavelets, each family is generated by a father wavelet or scaling function ϕ. Examples of such families are the *hat wavelets*, the *quadratic Battle-Lemarié wavelets*, and the *Shannon wavelets* (see nearby figures). Note that each of these functions belongs to $L^2(\mathbb{R})$ [10, 42]. (The spaces V_n that correspond to these respective wavelet families will be different than the Haar spaces. In particular, the compact support requirement is not necessary, as will be discussed in the next section.)

========================= **Problems** =========================

5. For each of the examples of wavelets just mentioned, determine the graphs of $\phi_{n,k}$ for your own choices of n and k.

6. Find a wavelet family different from those just mentioned. Be sure to cite your source. For the family you find, plot the graphs of $\phi_{1,0}$ and $\phi_{1,1}$.

7. Determine whether the quadratic Battle-Lemarié scaling function is a normalized scaling function.

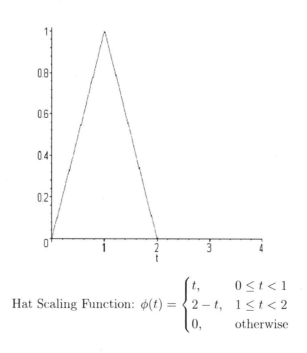

Hat Scaling Function: $\phi(t) = \begin{cases} t, & 0 \le t < 1 \\ 2 - t, & 1 \le t < 2 \\ 0, & \text{otherwise} \end{cases}$

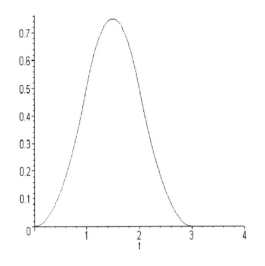

Quadratic Battle-Lemarié Scaling Function:

$$\phi(t) = \begin{cases} \frac{1}{2}t^2, & 0 \leq t < 1 \\ -t^2 + 3t - \frac{3}{2}, & 1 \leq t < 2 \\ \frac{1}{2}(t-3)^2, & 2 \leq t < 3 \\ 0, & \text{otherwise} \end{cases}$$

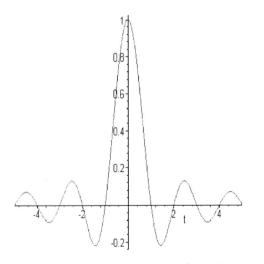

Shannon Scaling Function: $\phi(t) = \begin{cases} \frac{\sin(\pi t)}{\pi t}, & t \neq 0 \\ 1, & t = 0 \end{cases}$

At this point, other observations about these scaling functions should be noted.

- The Shannon scaling function is smooth. This means that the function and all of its derivatives exist and are continuous.

- Unlike the Haar scaling function, the Shannon function does not have compact support. The two other scaling functions do have compact support, and both are continuous. (In fact, the quadratic Battle-Lemarié scaling function has a continuous derivative.)

- For the hat and the quadratic Battle-Lemarié scaling functions, there exist integers k such that $\langle \phi(t), \phi(t-k) \rangle \neq 0$. The fact that each of these scaling functions is not orthogonal to all of its translates is a problem. We will discuss why in the next section.

Recall the relationship between the Haar father and mother wavelets, given in (1.1):

$$\psi(t) = \phi(2t) - \phi(2t - 1).$$

It is tempting to define the mother wavelet for each wavelet family by (1.1). However, the actual relationship between mother and father wavelets is unique to each family (not unlike real life!) and (1.1) only applies to the Haar family. This relationship will be explored later in this chapter.

Another wavelet family worthy of mention is the *Mexican hat family* (also called the *Maar wavelets*), which have been used in the study of underwater acoustics [14]. This family is different than the others encountered so far in that there is no simple form for the scaling function. There is, however, a

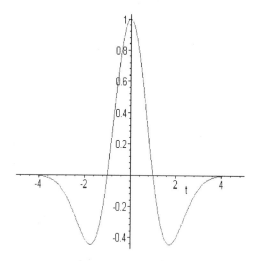

Mexican Hat Mother Wavelet: $\psi(t) = (1 - t^2)e^{(-t^2/2)}$

simple representation of the mother wavelet. Later in the text, it will be shown that the wavelet family can be generated from either the father or mother.

It is important to note that in *most* wavelet families, there is no simple form of either the father or mother. Instead, wavelets are defined by certain properties, and then approximated, as will be demonstrated in section 3.6.

======================== **Problems** ========================

8. Find a value of k so that $\langle \phi(t), \phi(t-k) \rangle \neq 0$ for each of the hat and the quadratic Battle-Lemarié examples.

9. A twice differentiable function whose second derivative is continuous is said to be a C^2 function. Is the quadratic Battle-Lemarié scaling function a C^2 function? Is its first derivative continuous?

10. Normalize the Mexican hat mother wavelet.

11. For the normalized Mexican hat family, determine graphs of

$$\psi_{n,k}(t) = \psi(2^n t - k)$$

for your own choices of n and k.

3.3 MULTIRESOLUTION ANALYSIS

It has been shown that wavelets are generated from a "parent" (father or mother) by *scalings* (contractions) and *translations* (horizontal shifts). When wavelets in V_n are used for negative values of n, signals can be analyzed on a large scale (over large intervals). By choosing positive values for n, signals can be isolated on a small scale (or interval). The beauty and power of wavelets is that, since there is an infinite collection of wavelets with which to analyze a signal, both of these tasks can be performed simultaneously.

We used the Haar wavelets in the previous chapters for their simplicity, a property that makes them ideal for demonstrating concepts. However, the Haar wavelets are not used in practice because they lack some important properties. Wavelets that are typically used in applications are constructed to satisfy certain criteria. The standard approach is to first build a *multiresolution analysis* (MRA) and then construct the wavelet family with the desired criteria from the MRA. In this section we discuss the concept of a multiresolution analysis.

Before an MRA can be defined, it will be helpful to discuss some additional properties of wavelets. The Haar wavelets will again illustrate these ideas.

The first property is called the *density property*. Density is a property that measures how intermingled the elements of one set are with another. For example, consider the rational numbers as a subset of the real numbers. For any real number, it is possible to find rational numbers arbitrarily close to that real number. In the example of the number e, the sequence $\{2, 2.7, 2.71, 2.718, 2.7182, \ldots\}$, obtained by truncating the decimal representation of e after each successive place, converges to e. In the same manner, a sequence of rational numbers that converges to any given real number always exists.

Another way to think about this is the following. Suppose there is a magnifier that can zoom in at any desired resolution to look at the real line. Wherever it zooms, no matter the magnification, there will always be both rational numbers and irrational numbers.

In general, a subset B of a set A is *dense* in A if any given element in A can be approximated as closely as we like by an element in B. The example above demonstrates how the set of rational numbers $\mathbb{Q}$ is dense in $\mathbb{R}$. An alternative characterization is that B is dense in A if, given any element $a \in A$, a sequence $\{b_n\}$ in B can be found that converges to a.

This idea can also be applied to function spaces. Recall that $L^2(\mathbb{R})$ is an inner product space with the inner product (3.1). In this setting, the distance between functions is measured by the norm of their differences. So the distance between the functions f and g in $L^2(\mathbb{R})$, denoted $d(f, g)$, is given by

$$d(f,g) = \|f - g\| = \left(\int_{-\infty}^{\infty} (f(t) - g(t))^2 dt \right)^{\frac{1}{2}}. \tag{3.2}$$

How does this apply to wavelets? In sections 2.4 and 2.5, we discussed the projection of functions onto wavelet spaces. Figures 3.1, 3.2, and 3.3 show the graph of the sine function on [0,1] and the projections of this function onto the Haar spaces V_2, V_4, and V_6. Notice that this function can be approximated as closely as we like by functions in V_n simply by choosing n as large as needed.

―――――――――――――――― **Problems** ――――――――――――――――

12. Let $f(t) = \sin(t)$ on [0,1]. Determine the projections of f onto the spaces V_2, V_4, and V_6. Compute the distance between f and its projection in each case using the integral norm (3.2). (Note: see problem 41 from chapter 2).

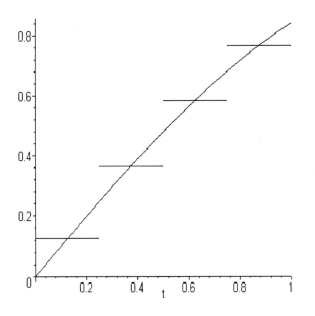

Fig. 3.1   Projection of $\sin(t)$ onto V_2.

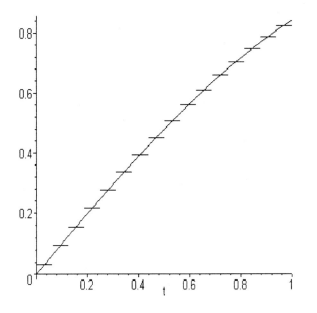

Fig. 3.2   Projection of $\sin(t)$ onto V_4.

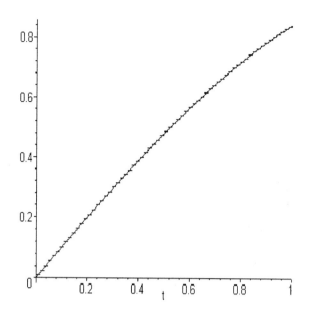

Fig. 3.3 Projection of $\sin(t)$ onto V_6.

In order to approximate any function in $L^2(\mathbb{R})$ as closely as we like by functions in the Haar spaces V_n, we need to have access to wavelets $\psi_{n,k}$ in V_n for arbitrarily large n. Rather than examine V_n for each n to determine the precision of an approximation, it is useful to consider the space $V = \bigcup_{n \in \mathbb{Z}} V_n$. An element in V, then, is a function that is piecewise constant on intervals of the form $\left[\frac{a}{2^m}, \frac{b}{2^m}\right]$ for some $m \in \mathbb{Z}$ and $a, b \in \mathbb{Z}$. In particular, V contains all the Haar wavelets. As the example with $f(t) = \sin(t)$ suggests, given any $f \in L^2(\mathbb{R})$, a sequence of functions (one in each V_n) can be constructed that converges to f. This means that $V = \bigcup_{n \in \mathbb{Z}} V_n$ is dense $L^2(\mathbb{R})$.

=========================== **Problems** ===========================

13. This exercise will lead to the following theorem: For the Haar wavelets, the set V_2 is not dense in $L^2(\mathbb{R})$.

    First note that if V_2 is dense in $L^2(\mathbb{R})$, then for every function $f \in L^2(\mathbb{R})$ and every $\epsilon > 0$, there exists a function $g \in V_2$ so that $\|f(t) - g(t)\| < \epsilon$.

    (a) Let f be the function defined by

    $$f(t) = \begin{cases} 2, & \text{if } 0 < t < \frac{1}{8} \\ 0 & \text{otherwise.} \end{cases}$$

    and let $\epsilon = \frac{3}{5}$. Find a function $g \in V_2$ so that $\|f(t) - g(t)\| < \epsilon$.

(b) Explain why you cannot find such a function g if $\epsilon = \frac{2}{5}$.

(c) Write the negation of the definition of density.

(d) Prove the theorem.

14. Prove that for Haar wavelets, the set V_3 is not dense in $L^2(\mathbb{R})$.

Now consider the set $I = \bigcap_{n \in \mathbb{Z}} V_n$. Recall that V_n consists of the functions that are piecewise constant on intervals of the form $\left[\frac{a}{2^m}, \frac{b}{2^m}\right]$. As n increases, the lengths of the intervals on which a function in V_n is constant approach zero. It follows that a function in I must be piecewise constant on every interval. The only functions that have this property are the constant functions, and the only constant function in $L^2(\mathbb{R})$ is the 0 function. This shows that

$$I = \bigcap_{n \in \mathbb{Z}} V_n = \{0\}.$$

Whenever we have a nested sequence

$$\cdots \subseteq V_{-1} \subseteq V_0 \subseteq V_1 \subseteq V_2 \subseteq \cdots$$

of sets satisfying $\bigcap_{n \in \mathbb{Z}} V_n = \{0\}$, the collection $\{V_n\}$ is said to have the *separation property*.

Problems

15. Verify that the only constant function in $L^2(\mathbb{R})$ is the zero function. Hint: Remember problem 1 of this chapter.

Next, let us examine how functions in the various sets V_n are related. Observe that if $f \in V_n$, then f is piecewise constant on intervals of length 2^{-n}. It follows that $f(2t)$ will be piecewise constant on intervals of length $2^{-(n+1)}$. This means that $f(2t)$ is in V_{n+1}. Similarly, $f\left(\frac{t}{2}\right) \in V_{n-1}$ and, by induction, $f\left(2^{-n}t\right) \in V_0$. Conversely, if $f\left(2^{-n}t\right) \in V_0$, then f must be in V_n.

We can use these general properties of the Haar wavelets to construct similar sequences of spaces that will contain other families of wavelets. This leads to the idea of a multiresolution analysis.

Definition: A *multiresolution analysis* (MRA) [10, 35] is a nested sequence

$$\cdots \subseteq V_{-1} \subseteq V_0 \subseteq V_1 \subseteq V_2 \subseteq \cdots$$

of subspaces of $L^2(\mathbb{R})$ with a scaling function ϕ such that

1. $\bigcup_{n \in \mathbb{Z}} V_n$ is dense in $L^2(\mathbb{R})$,

2. $\bigcap_{n \in \mathbb{Z}} V_n = \{0\}$,

3. $f(t) \in V_n$ if and only if $f(2^{-n}) \in V_0$, and

4. $\{\phi(t - k)\}_{k \in \mathbb{Z}}$ is an orthonormal basis for V_0.

Note that the fourth property makes it impossible for either the hat or the quadratic Battle-Lemarié scaling functions to form a multiresolution analysis.[1] Sometimes, though, a scaling function needs to merely be normalized in order to obtain a multiresolution analysis.

For most wavelets used in practice, there is no simple formula for the scaling function. Instead, a critical property of each scaling function follows from condition (4). Since $\{\phi(t - k)\}$ is an orthonormal basis for V_0, the set $\{\phi(2t - k)\}$ is an orthogonal basis for V_1. The fact that $\{\phi(2t - k)\}$ is a basis for V_1 means $\phi(t) \in V_0 \subset V_1$ can be written in the form[2]

$$\phi(t) = \sum_k c_k \phi(2t - k), \qquad (3.3)$$

for some constants c_k. This equation is called a *dilation equation* and is crucial in the theory of wavelets. (In some books and articles, it is referred to as a refinement equation or two-scale difference equation.) The constants $\{c_k\}$ are the *refinement coefficients*. The fact that a scaling function satisfies a dilation equation is a consequence of a multiresolution analysis. It will be shown that this equation provides enough information that we can proceed without knowing a specific formula for ϕ.

The dilation equation for the Haar wavelets

$$\phi(t) = \phi(2t) + \phi(2t - 1) \qquad (3.4)$$

was described in chapter 2. In this case, c_0 and c_1 are 1 and the rest of the refinement coefficients are 0.

Problems

16. Given a multiresolution analysis with scaling function ϕ, show that $\{\phi(2t - k)\}$ is an orthogonal basis for V_1.

For the next three problems, assume that with a multiresolution analysis every scaling function satisfies a dilation equation. Prove the following about refinement coefficients:

---

[1] There is a weaker version of multiresolution analysis, where condition (4) is replaced with a "Riesz basis" condition. For more information, see [10].

[2] Note that this could be an infinite sum and there are issues of convergence to be considered. However, we will deal mainly with situations where a finite number of the c_k are nonzero, and hence convergence is not a problem. (We will not deal with the general case.) This will allow us to interchange the order of sums and integrals when necessary.

17. $c_k = 2\langle \phi(t), \phi(2t - k)\rangle$.
   Hint: Use the Orthogonal Decomposition Theorem.

18. Parseval's formula: $\sum_{k=-\infty}^{\infty} c_k^2 = 2$
   Hint: Use the fact that $\langle \phi(t), \phi(t)\rangle = \langle \sum_k c_k \phi(2t - k), \sum_k c_k \phi(2t - k)\rangle$.

19. For every integer j, except zero, $\sum_{k=-\infty}^{\infty} c_k c_{k-2j} = 0$.
   Hint: Use (3.3) to write a dilation equation for $\phi(t - j)$.

3.4   THE HAAR SCALING FUNCTION REDISCOVERED

As mentioned in the previous section, when working with a multiresolution analysis we often don't have a simple formula for the scaling function. In these situations, however, we usually do know the refinement coefficients. Next, we will consider how such information enables us to determine some characteristics of the scaling function.

One of the standard techniques to deal with problems like this is a numerical one known as the *cascade algorithm* [10]. This algorithm will provide approximations to the scaling function and is an example of a *fixed-point method*.

A *fixed point* of a function f is a value a such that $f(a) = a$. A simple example of a fixed point method follows, which shows how to determine the fixed point of the cosine function. Let $f(t) = \cos(t)$. To find a solution to the equation $\cos(t) = t$, start with a guess, t_0, of a fixed point. Let $t_1 = f(t_0) = \cos(t_0)$, then compute another number $t_2 = f(t_1) = \cos(t_1)$, and continue. In this way, construct a sequence

$$\{t_0, t_1 = f(t_0), t_2 = f(t_1), \ldots, t_{n+1} = f(t_n), \ldots\}$$

that will converge to the fixed point of the cosine function. This process will work for functions f that satisfy certain conditions.

===================== **Problems** =====================

20. Use the algorithm to determine, to six decimal places of accuracy, the solution of

$$t = \cos(t).$$

21. Use the algorithm to determine, to five decimal places of accuracy, the solution of

$$t = 1 + e^{-t}.$$

22. Explain what happens when you apply the algorithm to attempt to determine the solution of

$$t = 3.7t(1 - t) + 0.2.$$

How is this function different than the other two?

The cascade algorithm is a fixed-point method, except that instead of generating a sequence of numbers, it creates a sequence of functions. When given refinement coefficients, this algorithm creates a sequence of functions $\{f_i\}$ so that, for every value of t, $f_i(t) \to \phi(t)$ as $i \to \infty$. Recall that ϕ satisfies the dilation equation (3.3). If we let F be the function that assigns the expression

$$F(\gamma)(t) = \sum_n c_n \gamma(2t - n)$$

to any function γ, then we can consider ϕ as a fixed point of F! This process will be illustrated using the Haar dilation equation (3.4).

Begin with a guess, $f_0(t)$, of the graph of $\phi(t)$ (imagining for a moment that we don't know the Haar scaling function). Every scaling function that we have seen so far has a maximum near $t = 0$ and tends to get smaller as we move away from that maximum. So, a good first guess for $f_0(t)$ could be the normalized tent function,

$$f_0(t) = \begin{cases} 1 + t, & \text{if } -1 \le t < 0 \\ 1 - t, & \text{if } 0 \le t < 1 \\ 0, & \text{otherwise.} \end{cases}$$

_____ **Problems** _____

23. Graph $f_0(t)$.

There is an alternate way to define $f_0(t)$ that illuminates how the cascade algorithm works. Think of creating $f_0(t)$ as a three-step process reminiscent of the trapezoid rule from calculus. First, divide the t-axis into subintervals with breaks at each integer. Second, give the function a value at each of the integers. In this case, it is zero at every integer except at 0, where the function is equal to 1. Finally, use linear segments to connect the function values on the integers. For the current f_0 we use $1 + t$ on the interval from 1 to 0, $1 - t$ on the interval from 0 to 1, and zero everywhere else. The resulting function is called a _linear spline_.

_____ **Problems** _____

24. Use a CAS to graph $f_0(t)$ in the following way. First, define a list of coordinates, based on the values of f_0 at the integers. (You do not need all of the integers. Focus on the interval $[-1, 4]$.) Then, plot these points, connecting them with lines.

Maple **Hint:** Using the `style=line` option with the `plot` command will draw straight lines between the points.

We now use $f_0(t)$ to create a new and better approximation $f_1(t)$. (Daubechies calls this step "cranking the machine".) From (3.4) and the fact that $f_1 = F(f_0)$, it follows that

$$f_1(t) = f_0(2t) + f_0(2t - 1).$$

From here, use a three step process: *update, extend,* and *connect* to more fully define f_1. First, find the value of f_1 on the integers. For example,

$$f_1(1) = f_0(2) + f_0(1) = 0 + 0 = 0.$$

Repeating this for each integer shows that f_1, like f_0, is equal to zero on all of the integers except 0, where it is equal to 1. Not terribly impressive yet, but this is just the first step.

Now that we have values for f_1 at the integers, extend the function to the points halfway between the integers. For example,

$$f_1\left(\frac{1}{2}\right) = f_0(1) + f_0(0) = 0 + 1 = 1.$$

Calculations like this one show that f_1 is zero at the odd multiples of $\frac{1}{2}$ with the single exception that $f_1(\frac{1}{2}) = 1$. Finally, complete the definition of f_1 by using linear segments to connect the function values on the integers and the halves, as was done with f_0.

Problems

25. Graph f_1 using the method of exercise 24.

Now that f_1 has been determined, we can then "crank the machine" again to determine f_2. This time, first update the values on the set

$$\left\{ \ldots, \ -1, \ -\frac{1}{2}, \ 0, \ \frac{1}{2}, \ 1, \ \ldots \right\},$$

and then extend to the odd multiples of $\frac{1}{4}$. Finally, connect the function values on all of the multiples of $\frac{1}{4}$.

Problems

26. Find f_2 and sketch its graph. Note that each graph that you have been generating looks more and more like the Haar scaling function with which we are familiar.

27. "Cranking the machine" one step at a time is time consuming and tiresome. Create a CAS worksheet with loops that will perform the cascade algorithm and plot the graph of f_8. How similar are f_8 and the Haar scaling function?

Maple **Hint:** One can use either the symbolic or numeric power of *Maple* to generate f_8. To force *Maple* to use floating-point arithmetic, rather than symbolic algebra, use a decimal point when defining the refinement coefficients. For example,

```
> c[0]:=1.0;
```

Here is one way to create a loop (in *Maple*) to perform this pointwise algorithm. Begin with the refinement coefficients. Here, we use the Haar coefficients

```
> c[0]:=1.0; > c[1]:=1.0;
```

Label the ith approximation of ϕ as f[i]. We will plot these approximations on the interval $[-1, 4]$. Remember, at each step in the algorithm,

$$f[i+1](t) = c[0]*f[i](2*t) + c[1]*f[i](2*t-1).$$

Note that to define an approximation on $[-1, 4]$, we need values for the prior approximation from -3 to 8. First define f[0] at integer points from -3 to 8.

```
> for i from -3 to 8 do f[0](i):=0.: od:   f[0](0):=1.:
```

To plot this approximation we construct a list of points, called points[0], and connect them with line segments.

```
> points[0]:=[[ t, f[0](t)] $t=-1..4];
> plot(points[0],style=line);
```

The following loop generates successive pointwise approximations to the scaling function ϕ. The first approximation is defined on the halves. The second on the quarters, and so on. All of these approximations are restricted to $[-1, 4]$.

```
> for j from 1 to 8 do deltak := 2^(-j):
> for k from 0 to 11/deltak do
> x:=-3+k*deltak:
> if -1<=x and x<=4 then
> f[j](x) := c[0]*f[j-1](2*x) + c[1]*f[j-1](2*x-1):
```

```
> else f[j](x):=0: fi:
> od:
> points[j] := [[t*2^(-j),f[j](t*2^(-j))]$t=-1*2^j..4*2^j]:
> od:
```

Now `points[j]` is the jth approximation. The successive approximations can be animated in *Maple*; the `plots` package is needed to do this.

```
> with(plots):
> for j from 0 to 8 do
> myplot[j]:=plot(points[j], t=-1..4, style=line,axes=box):
> od:
> display(seq(myplot[i], i=0..8), insequence=true);
```

28. After how many iterations does it make sense to stop "cranking the machine"? Justify your answer.

Many of the examples of wavelet families from section 3.2 have simple refinement coefficients [38]. The scaling function for the hat wavelet satisfies the dilation equation

$$\phi(t) = \frac{1}{2}\phi(2t) + \phi(2t-1) + \frac{1}{2}\phi(2t-2),$$

while the quadratic Battle-Lemarié scaling function satisfies

$$\phi(t) = \frac{1}{4}\phi(2t) + \frac{3}{4}\phi(2t-1) + \frac{3}{4}\phi(2t-2) + \frac{1}{4}\phi(2t-3). \qquad (3.5)$$

Problems

29. Modify your CAS worksheet from problem 27 to approximate the hat wavelet scaling function.

30. Modify your CAS worksheet to approximate the quadratic Battle-Lemarié wavelet scaling function. (Note: the support of this scaling function is $[0,3]$.)

31. The cubic Battle-Lemarié wavelet scaling function satisfies the dilation equation

$$\phi(t) = \frac{1}{8}\phi(2t) + \frac{1}{2}\phi(2t-1) + \frac{3}{4}\phi(2t-2) + \frac{1}{2}\phi(2t-3) + \frac{1}{8}\phi(2t-4).$$

Modify your CAS worksheet to approximate the cubic Battle-Lemarié wavelet scaling function. (Note: the support of this scaling function is $[0,4]$.)

32. Determine which, if any, of the functions in problems 29 through 31 are normalized.

33. Modify your CAS worksheet to begin with the function f_0 defined by

$$f_0(t) = \begin{cases} 2(1+t), & \text{if } -1 \le t < 0 \\ 2(1-t), & \text{if } 0 \le t < 1 \\ 0, & \text{otherwise.} \end{cases}$$

(Note that this f_0 is twice the tent function and is not normalized.) How does this change affect your results in problems 29 through 31? What conjecture can you draw about the cascade algorithm and normalization? Can you prove your conjecture?

Both Battle-Lemarié wavelet scaling functions are examples of what are called *bell-shaped splines*, or simply *B*-splines. A spline is a function where several polynomials, defined on different sub-intervals, are joined together to create a continuous function. For example, the quadratic Battle-Lemarié scaling function is created by joining three different quadratic functions together, along with the zero function. A project featuring *B*-splines can be found in chapter 4.

====================== **Problems** ======================

34. (a) Explain why the hat scaling function could be called the "Linear Battle-Lemarié scaling function." What about the Haar scaling function?

    (b) The space C^q consists of all functions whose q^{th} derivative is continuous. For each of the four scaling functions (Haar, hat, quadratic Battle-Lemarié, cubic Battle-Lemarié), determine the value of q so that it is correct to say that the function is a C^q function. What is the pattern?

35. (a) Show that the function

$$\phi(t) = \begin{cases} \frac{3+t}{3}, & \text{if } -3 \le t \le 0 \\ \frac{3-t}{3}, & \text{if } 0 < t \le 3 \\ 0, & \text{otherwise} \end{cases}$$

    is a solution to the dilation equation

$$\phi(t) = \frac{1}{2}\phi(2t-3) + \phi(2t) + \frac{1}{2}\phi(2t+3).$$

(b) Modify your CAS worksheet to explore what happens when you try to solve this dilation equation [12]. Explain what happens and the relationship to problem 22 of this chapter.

36. Start with a function different than the tent function, and use the cascade algorithm to try to create the quadratic Battle-Lemarié wavelet scaling function. What is the result? A function, f, that leads to some neat pictures is

$$f(t) = \begin{cases} \frac{1}{2}, & \text{if } -1 < t < 1 \\ 0, & \text{otherwise.} \end{cases}$$

Another approach to find the scaling function ϕ, introduced by Strang, uses matrices [34]. For example, suppose we start with the dilation equation for the quadratic Battle-Lemarié scaling function (3.5) and *assume* that the support of $\phi(t)$ is $0 \leq t \leq 3$ and $\phi(0) = 0 = \phi(3)$. Substituting $t = 1$ and $t = 2$ into (3.5), gives us the following two equations:

$$\phi(1) = \frac{1}{4}\phi(2) + \frac{3}{4}\phi(1)$$

$$\phi(2) = \frac{3}{4}\phi(2) + \frac{1}{4}\phi(1).$$

These equations can be viewed in matrix form as $\mathbf{x} = L\mathbf{x}$, where

$$\mathbf{x} = \begin{bmatrix} \phi(1) \\ \phi(2) \end{bmatrix} \quad \text{and} \quad L = \begin{bmatrix} 3/4 & 1/4 \\ 1/4 & 3/4 \end{bmatrix}.$$

Any solution $\mathbf{x}$ of this linear equation will be an eigenvector of L with eigenvalue 1. In this case it turns out that $\phi(1) = \phi(2)$. After choosing an arbitrary value for $\phi(1)$, and hence $\phi(2)$, we can then use the dilation equation to determine the values of ϕ on the halves, quarters, etc., in a fashion similar to the cascade algorithm. Finally, all values are multiplied by an appropriate constant so that the norm of ϕ is 1.

Problems

37. Apply Strang's approach to the dilation equation for the cubic Battle-Lemarié scaling function to determine the relationship between the function values at 1, 2, and 3.

3.5   RELATIONSHIPS BETWEEN THE MOTHER
AND FATHER WAVELETS

In the previous chapters we worked with wavelets, $\psi_{n,k}$, that were generated from the mother wavelet ψ. Recall that $\psi_{n,k}(t) = \psi(2^n t - k)$ for k from 0 to $2^n - 1$. However, in many instances we know the father wavelet, or scaling function, but do not have a specific formula for the mother wavelet. In (1.1) we saw how the Haar mother wavelet could be written in terms of the scaling function. In this section we will see that this is true in general, provided we have a multiresolution analysis. This will enable us to work with wavelets, even if all we have is a scaling function.

Assume that we have a multiresolution analysis with a scaling function (father wavelet). In chapter 2, we computed the wavelet coefficients for the Haar wavelets by calculating the projections of signals onto the spaces V_n and $V_n^{\perp}$. In fact, the Orthogonal Decomposition Theorem guarantees that each signal has a unique decomposition in $V_n \oplus V_n^{\perp}$. Specifically, the projection of a signal or function $f \in L^2(\mathbb{R})$ onto $V_n^{\perp}$ was given by

$$\sum_k \frac{\langle f, \psi_{n,k} \rangle}{\langle \psi_{n,k}, \psi_{n,k} \rangle} \psi_{n,k}.$$

Let $P_n(f)$ denote the projection of f onto V_n. Since $V_{n+1} = V_n \oplus V_n^{\perp}$, it follows that, for the Haar wavelets, one can write

$$P_{n+1}(f) = P_n(f) + \sum_k \frac{\langle f, \psi_{n,k} \rangle}{\langle \psi_{n,k}, \psi_{n,k} \rangle} \psi_{n,k}. \qquad (3.6)$$

The theory of multiresolution analyses states that *whenever* we have an MRA, there is always a function ψ that generates an orthonormal wavelet basis

$$\left\{ \psi_{n,k} = 2^{-k/2} \psi\left(2^n t - k\right), 0 \le k \le 2^n - 1, n \in \mathbb{Z} \right\}$$

of $L^2(\mathbb{R})$ so that (3.6) holds for any $f \in L^2(\mathbb{R})$. (How this arises from the Orthogonal Decomposition Theorem was demonstrated in Chapter 2.) The factor of $2^{-k/2}$ that appears in $\psi_{n,k}(t)$ normalizes these functions.

How do we relate ψ to the scaling function ϕ? Given an MRA with scaling function ϕ, there is a dilation equation

$$\phi(t) = \sum_k c_k \phi(2t - k).$$

Note that since ϕ is a solution of this equation, so is any scalar multiple of ϕ. This means that ϕ may be normalized (just multiply by a suitable constant). In other words, it can be assumed that

$$\langle \phi(t), \phi(t) \rangle = \int_{-\infty}^{\infty} \phi^2(t) \, dt = 1.$$

In addition, with an MRA, $\{\phi(t - k)\}_{k \in \mathbb{Z}}$ is an orthonormal set. A straightforward substitution shows that $\{\phi(2t - k)\}$ is also an orthogonal set. In other words, $\langle\phi(2t - k), \phi(2t - m)\rangle = 0$ for $k \neq m$. The reader is encouraged to verify this fact in the problems that follow.

The fact that $\{\phi(2t - k)\}$ is orthogonal also implies that $\{\phi(2t - k)\}$ is a linearly independent set (problem 39). Note, however, that

$$\langle\phi(2t - k), \phi(2t - k)\rangle = \int_{-\infty}^{\infty} \phi^2(2t - k) \, dt = \frac{1}{2} \int_{-\infty}^{\infty} \phi^2(u) \, du = \frac{1}{2}, \quad (3.7)$$

so $\{\phi(2t - k)\}$ is not an orthonormal set.

To make computations easier, normalize the functions $\phi(2t - k)$ by multiplying each by $\sqrt{2}$. This gives a new dilation equation with *normalized* refinement coefficients $h_k = \frac{c_k}{\sqrt{2}}$ so that

$$\phi(t) = \sum_k c_k \phi(2t - k) = \sum_k h_k \sqrt{2}\phi(2t - k). \quad (3.8)$$

Problems

38. Verify that $\langle\phi(2t - k), \phi(2t - m)\rangle = 0$ for $k \neq m$.
    Hint: Use the substitution $u = 2t$.

39. Show that if $\{v_1, v_2, \ldots, v_n\}$ is an orthogonal set that does not contain 0 in an inner product space V, then $\{v_1, v_2, \ldots, v_n\}$ is a linearly independent set.

40. Show that $\sum_k h_k^2 = 1$.
    Hint: Use problem 18 of this chapter and the fact that $c_k = \sqrt{2}h_k$ from (3.8).

Continuing with the problem of relating the mother wavelet to the father, recall that the mother Haar wavelet could be written in terms of the father wavelet by

$$\psi(t) = \phi(2t) - \phi(2t - 1).$$

This equation can be written in the more general form

$$\psi(t) = \sum_k g_k \sqrt{2}\phi(2t - k),$$

with $g_0 = \frac{1}{\sqrt{2}}$, $g_1 = -\frac{1}{\sqrt{2}}$ and $g_i = 0$ for all other i. In the Haar case the father wavelet has the form

$$\phi(t) = \phi(2t) + \phi(2t - 1).$$

Comparing this to (3.8), we find that $h_0 = \frac{1}{\sqrt{2}} = h_1$, and $h_i = 0$ for all other i. So, for the Haar wavelets, $g_0 = h_1$, $g_1 = -h_0$, and $g_i = 0$ for all other i. (Of course, we could have $g_0 = h_0$, $g_1 = -h_1$, but we will see later why these particular assignments were chosen.)

Let us assume, like with the Haar wavelets, that an MRA implies that ψ has the form

$$\psi(t) = \sum_k g_k \sqrt{2}\phi(2t - k).$$

What does multiresolution theory tell us about the coefficients g_k?

Using (3.7) and problem 38 we obtain

$$\langle \phi(t), \psi(t) \rangle = \left\langle \sum_k h_k \sqrt{2}\phi(2t - k), \sum_m g_m \sqrt{2}\phi(2t - m) \right\rangle$$

$$= \sum_k h_k \left\langle \sqrt{2}\phi(2t - k), \sum_m g_m \sqrt{2}\phi(2t - m) \right\rangle$$

$$= \sum_k h_k \sum_m g_m \left\langle \sqrt{2}\phi(2t - k), \sqrt{2}\phi(2t - m) \right\rangle$$

$$= \sum_{k,m} h_k g_m \left\langle \sqrt{2}\phi(2t - k), \sqrt{2}\phi(2t - m) \right\rangle$$

$$= \sum_k h_k g_k \left\langle \sqrt{2}\phi(2t - k), \sqrt{2}\phi(2t - k) \right\rangle$$

$$+ \sum_{k \neq m} h_k g_m \left\langle \sqrt{2}\phi(2t - k), \sqrt{2}\phi(2t - m) \right\rangle$$

$$= \sum_k 2 h_k g_k \langle \phi(2t - k), \phi(2t - k) \rangle$$

$$= \sum_k h_k g_k.$$

If we want the mother and father wavelets to be orthogonal (recall our discussion in chapter 2, where $\phi \in V_0$ and $\psi \in V_0^{\perp}$), then the preceeding string of equalities implies

$$\sum_k h_k g_k = 0. \tag{3.9}$$

In addition, for $k, m \in \mathbb{Z}$,

$$\langle \psi(t - k), \psi(t - m) \rangle = \left\langle \sum_i g_i \sqrt{2}\phi(2t - 2k - i), \sum_j g_j \sqrt{2}\phi(2t - 2m - j) \right\rangle$$

$$= \sum_{i,j} 2 g_i g_j \langle \phi(2t - (i + 2k)), \phi(2t - (j + m)) \rangle.$$

Note that all terms disappear except for those where $i + 2k = j + 2m$. So

$$\langle \psi(t-k), \psi(t-m)\rangle = \sum_{i+2k=j+2m=q} 2g_i g_j \langle \phi(2t-q), \phi(2t-q)\rangle$$

$$= \sum_i g_i g_{i-2(k-m)}.$$

It is a straightforward exercise to show that $\{\psi(t-k)\}_{k \in \mathbb{Z}}$ is an orthonormal basis for $V_0^\perp$. From this and the most recent chain of equalities we see that

$$\sum_i g_i g_{i-2(k-m)} = \begin{cases} 0, & \text{if } k \neq m \\ 1, & \text{if } k = m. \end{cases} \tag{3.10}$$

There are many possible sequences $\{g_k\}$ that satisfy (3.9) and (3.10). The values generally accepted for g_k are given by[3]

$$g_k = (-1)^k h_{1-k}. \tag{3.11}$$

In this case, the mother wavelet is given by

$$\psi(t) = \sum_k g_k \sqrt{2}\phi(2t - k) = \sum_k (-1)^k h_{1-k} \sqrt{2}\phi(2t - k). \tag{3.12}$$

_____ **Problems** _____

In this section we began with a multiresolution analysis, assumed ψ could be written in the form $\psi(t) = \sum_k g_k \sqrt{2}\phi(2t - k)$, and ultimately arrived at (3.11). In the problems below, we argue in the opposite direction. In other words, assume ψ is defined by (3.12).

41. Show that ψ is orthogonal to ϕ. (Hint: Arrive at (3.9) and use (3.11) to see how to split the sum into two appropriate pieces.)

42. (a) Show that $\{\psi(t - k)\}$ is an orthonormal basis for $V_0^\perp$.

    (b) Show that (3.10) holds. (Hint: Compare problem 19.)

43. For the Haar wavelets, show that $\sum_k g_k = 0$.

_____

[3] Although we have attempted to avoid references to Fourier analysis throughout this book, the close relationship between Fourier analysis and wavelet theory makes that task rather challenging. The relationship (3.11) is generally derived through Fourier analysis using orthogonality. There is a degree of freedom in defining these coefficients, and some authors use the definition $g_k = (-1)^{1-k} h_{1-k}$. The details may be found in [10] and [30].

3.6   DAUBECHIES WAVELETS

So far we have seen a variety of scaling functions. Of these, the Haar and Shannon scaling functions are, in a sense, at two extremes. The Haar father wavelet has compact support, but is discontinuous. On the other hand, the Shannon scaling function is smooth — all of its derivatives exist and are continuous — but its support is all of $\mathbb{R}$. Since these wavelet families are at the far ends of the support and continuity spectra, neither is ideal for use in applications. Rather, some sort of compromise between compact support and smoothness is needed, and one was discovered by Ingrid Daubechies in 1987 [25].

Daubechies sought a wavelet family that had compact support and some sort of smoothness. Starting with certain explicit requirements on the wavelets, she determined the appropriate refinement coefficients, and, using the cascade algorithm, developed a graph of a scaling function. Her discovery that one could actually find a scaling function, given the conditions she stated, was quite a feat, and was greeted with enthusiasm [19]. Daubechies actually developed a number of related wavelet families, and we will now consider one of the simpler examples.

There are three requirements for the following example of Daubechies wavelets. The first condition is that the scaling function has compact support, in particular, that $\phi(t)$ is zero outside of the interval $0 < t < 3$. A consequence of this is that all the refinement coefficients are zero except c_0, c_1, c_2 and c_3. Note that this implies $\phi(t) = c_0\phi(2t) + c_1\phi(2t-1) + c_2\phi(2t-2) + c_3\phi(2t-3)$. This requirement is called the *compact support condition*.

═══════════════ Problems ═══════════════

44. Use the dilation equation (3.3) to prove that if $\phi(t)$ is zero outside of the interval $0 < t < 3$, then $c_n = 0$ if n is not 0, 1, 2, or 3. (Hint: using problem 17, see what happens when you use the following values of k: -4, -3, 6, and 7. Then, observe what happens when you substitute the following values of t into the dilation equation: $-\frac{1}{2}$, 3.)

The second requirement satisfied by this class of wavelets is the *orthogonality condition*, which is at the heart of any multiresolution analysis. Simply stated, this condition requires that the scaling function be orthogonal to its translates. As seen in exercises 18 and 19, the orthogonality condition implies that $\sum_k c_k^2 = 2$ and $\sum_k c_k c_{k-2m} = 0$ for any m. When applied to this class of Daubechies wavelets we get

$$c_0^2 + c_1^2 + c_2^2 + c_3^2 = 2, \tag{3.13}$$

and, using $m = 1$,

$$c_0 c_2 + c_1 c_3 = 0. \tag{3.14}$$

Problems

45. Why are other values of m not used in (3.14)?

46. It can be shown (through Fourier analysis) that one consequence of a multiresolution analysis is that

$$\int_{-\infty}^{\infty} \phi(t)dt \neq 0.$$

Use this fact and the dilation equation to prove that $c_0 + c_1 + c_2 + c_3 = 2$.

The final condition on these Daubechies wavelets is called the *regularity condition*, and it is related to the smoothness of the scaling function. The basic idea is that the smoother the scaling function is, the better the wavelet family can approximate polynomials. In this example, we want to ensure that all constant and linear polynomials can be written as a linear combination of elements in $\{\phi(t - k)\}$ [10]. (This may need to be an infinite sum.)

Two simple constant and linear functions are the functions 1 and t. Although these functions are not members of $L^2(\mathbb{R})$, in a sense, the requirement above makes them "honorary members" of V_0. Then, using ideas from chapter 2, these functions ought to be orthogonal to the mother wavelet ψ. This leads to the following two equations[4] (often referred to as *moment conditions*):

$$\int_{-\infty}^{\infty} \psi(t) \, dt = 0 \quad \text{and} \quad \int_{-\infty}^{\infty} t\psi(t) \, dt = 0.$$

It is also said, under this condition, that the mother wavelet has *vanishing moments*.

These moment conditions lead to two more equations that involve the refinement coefficients. Using the fact that the mother wavelet is defined in terms of the father wavelet and the refinement coefficients we obtain

$$\psi(t) = \sum_{-\infty}^{\infty} (-1)^k c_{1-k} \phi(2t - k).$$

Keeping in mind that, for the Daubechies refinement coefficients, only c_0, c_1, c_2, and c_3 are nonzero, it follows that

$$\psi(t) = -c_0 \phi(2t) + c_1 \phi(2t - 1) - c_2 \phi(2t - 2) + c_3 \phi(2t - 3). \qquad (3.15)$$

The subsequent exercises will show how the moment conditions lead to the following two equations:

$$-c_0 + c_1 - c_2 + c_3 = 0 \qquad (3.16)$$
$$-c_1 + 2c_2 - 3c_3 = 0. \qquad (3.17)$$

---

[4]A formal derivation of these equations makes extensive use of Fourier analysis.

_____ **Problems** _____

47. Derive equations (3.16) and (3.17) in the following fashion:

    (a) Show that, for any k,

    $$\int_{-\infty}^{\infty} \phi(2t - k) \, dt = \frac{1}{2} \int_{-\infty}^{\infty} \phi(t) \, dt.$$

    (Hint: Use an appropriate substitution.)

    (b) Show that, for any k,

    $$\int_{-\infty}^{\infty} t\phi(2t - k) \, dt = \frac{k}{4} \int_{-\infty}^{\infty} \phi(t) \, dt + \frac{1}{4} \int_{-\infty}^{\infty} t\phi(t) \, dt.$$

    (c) Use (a), the first moment condition, problem 46, and (3.15) to verify (3.16).

    (d) Use the second moment condition, problem 46, (3.15), (3.16), and (b) to prove (3.17) holds.

48. The conditions (3.13), (3.14), (3.15), (3.16), and problem 46 give us five equations in four unknowns. Show that

$$c_0 = \frac{1 + \sqrt{3}}{4}, \quad c_1 = \frac{3 + \sqrt{3}}{4}, \quad c_2 = \frac{3 - \sqrt{3}}{4}, \quad c_3 = \frac{1 - \sqrt{3}}{4}$$

is one solution to this system.

═══

Using the refinement coefficients from problem 48, we get the dilation equation

$$\phi(t) = \frac{1 + \sqrt{3}}{4}\phi(2t) + \frac{3 + \sqrt{3}}{4}\phi(2t - 1) + \frac{3 - \sqrt{3}}{4}\phi(2t - 2) + \frac{1 - \sqrt{3}}{4}\phi(2t - 3).$$

The scaling function that arises from this dilation equation is referred to as D_4, since there are four refinement coefficients.

_____ **Problems** _____

49. (a) Modify your CAS worksheet from the previous section to approximate D_4 via the cascade algorithm. Compare your result to the function on the cover of this book.

    (b) There are two solutions that can be obtained in problem 48. Find the second one. Use the cascade algorithm to graph the scaling functions for this second set of refinement coefficients. How are the scaling functions related? Is it fair to say that the two solutions give "essentially" the same solution? Explain.

50. For the D_4 wavelets, show that $\sum_k g_k = 0$ is satisfied.

51. Use a CAS to draw an approximation of the graph of ψ of the D_4 wavelets. (Hint: Approximate ϕ as a function (a *Maple* routine is included in the appendix), then use (3.12)).

It may seem surprising that Daubechies' three conditions would lead to such an odd-looking scaling function D_4. Daubechies herself has said, "I guess we just have to live with the way they look" [40]. It is not difficult to see that D_4 has compact support, but what about the other two conditions?

Problems

52. (Orthogonality) Use a CAS and some ideas from Riemann sum approximations to investigate numerically whether or not $\langle D_4(t), D_4(t-k) \rangle = 0$ for three different nonzero values of k. In a similar vein, investigate whether or not $\langle D_4(t), D_4(t) \rangle = 1$.

Although the regularity condition is difficult to observe, it turns out to be true that D_4 is a continuous function, although it is not differentiable everywhere. Therefore, D_4 belongs to C^0, but not C^1.

Problems

53. Suppose we wish to invent a scaling function with the following properties.

    Compact support: the support of the scaling function is $0 < t < 1$.

    Orthogonality: the identities from exercises 18 and 19 hold.

    Regularity: constant functions can be written as a linear combination of elements in $\{\phi(t-k)\}$.

    Determine the refinement coefficients for this scaling function. Have we seen this wavelet family before?

54. Apply Strang's idea from section 3.4 to the dilation equation for D_4 to determine the relationship between the function values at 1 and 2.

55. Although it is not a scaling function, the de Rham function [12] is interesting because it is a continuous function that is differentiable nowhere on $\mathbb{R}$. It satisfies the following dilation equation, which uses a coefficient of 3 rather than 2:

$$R(t) = \frac{2}{3}R(3t+2) + \frac{1}{3}R(3t+1) + R(3t) + \frac{1}{3}R(3t-1) + \frac{2}{3}R(3t-2).$$

    Modify your CAS worksheet from the previous section to create a graph of R using the cascade algorithm.

56. A variation on the derivation of the refinement coefficients for D_4 is to not use the moment conditions [6]. This leads to three equations in four unknowns, so the solutions can be defined in terms of a parameter θ. They are:

$$c_0 = \frac{1}{2}(1 - \cos(\theta) + \sin(\theta)) \quad c_1 = \frac{1}{2}(1 + \cos(\theta) + \sin(\theta))$$
$$c_2 = \frac{1}{2}(1 + \cos(\theta) - \sin(\theta)) \quad c_3 = \frac{1}{2}(1 - \cos(\theta) - \sin(\theta)).$$

Which angle θ gives the refinement coefficients for D_4? Which angle θ gives you the refinement coefficients for the Haar wavelets? Choose a different value for θ and construct the scaling function that arises. Then, create an animation on a CAS that shows how ϕ will change as you vary θ.

We have seen some important ideas in this section. For the elementary Daubechies wavelets, the compact support, orthoganality, and regularity conditions are all that are needed to construct the scaling function to whatever accuracy we desire. From this we can approximate the mother wavelet and all children as well. This indicates that it is not necessary to specifically know a formula for the scaling function in order to work with wavelets. Other wavelet families are developed in this way. See chapter 4 for a project investigating the Daubechies scaling function D_6.

3.7   HIGH AND LOW PASS FILTERS

Now that we have seen more families of wavelets, including wavelets with no simple algebraic representation, we return to a discussion of how they may be used to process signals. Given a multiresolution analysis, we can employ what is referred to as a "pyramidal" algorithm for processing. This algorithm depends on two sequences, called *filters*.

Recall from section 3.5 that the father and mother Haar wavelets satisfy the equations $\phi(t) = \phi(2t) + \phi(2t - 1)$ and $\psi(t) = \phi(2t) - \phi(2t - 1)$. More generally, for any wavelet family there will always be equations of the form

$$\phi(t) = \sum_k h_k \sqrt{2}\phi(2t - k) \quad \text{and} \quad \psi(t) = \sum_k g_k \sqrt{2}\phi(2t - k) \qquad (3.18)$$

for the father and mother wavelets.

The sequences $\{h_k\}$ and $\{g_k\}$ arising from these scaling equations must satisfy certain conditions, some of which have been explored in previous exercises. The most useful for our purposes is (3.11), which is again given here

$$g_k = (-1)^k h_{1-k}.$$

Remember that equation (3.11) allows us to construct the mother wavelet if we know the refinement coefficients and the scaling function. Even if we don't explicitly know the scaling function, the refinement coefficients and the cascade algorithm can be used to approximate ϕ and then ψ by (3.11) and (3.18).

Problems

57. Even though the hat scaling function does not define a family of orthogonal wavelets, draw the graph of ψ defined by (3.12).

58. Repeat problem 57 for the quadratic Battle-Lemarié scaling function.

59. Draw an approximation to the graph of ψ, as defined by (3.12), for the cubic Battle-Lemarié wavelets.

The sequences $\{h_k\}$ and $\{g_k\}$ that appear in (3.18) can be used to process signals and are typically called *low pass* and *high pass* filters, respectively.[5]
Let
$$s = [s_0, s_1, \ldots, s_{m-1}]$$
be a signal[6] of length 2^n for some n. Recall that this signal defines a function $f \in V_n$ given by
$$f = \sum_{k=0}^{m-1} s_k \phi_{n,k}.$$
The filters process signals by defining two *operators* H and G that, when applied to a signal s produce two new signals, Hs and Gs, each half the length of s. The kth entries of the new signals Hs and Gs, denoted by $(Hs)_k$ and $(Gs)_k$, respectively, are defined as
$$(Hs)_k = \sum_{j=0}^{2^n-1} h_{j-2k} s_j \tag{3.19}$$
and
$$(Gs)_k = \sum_{j=0}^{2^n-1} g_{j-2k} s_j. \tag{3.20}$$

While these operators are defined for any wavelet family, it is instructive to analyze them for the Haar wavelets. In this case $h_0 = h_1 = \frac{1}{\sqrt{2}}$ and $h_i = 0$

---

[5] It would seem that the high pass filter should be $\{h_k\}$, but that just isn't the case in the literature.
[6] For ease of reading, we will not use the "T" for transpose in this section.

for all other i. From (3.11), $g_0 = h_1 = \frac{1}{\sqrt{2}}$, $g_1 = -h_0 = -\frac{1}{\sqrt{2}}$, and $g_i = 0$ for all other i. So, if $\mathbf{s} = [1, 2, 3, -1, 1, -4, -2, 4]$, then

$$H\mathbf{s} = [h_0 s_0 + h_1 s_1, h_0 s_2 + h_1 s_3, h_0 s_4 + h_1 s_5, h_0 s_6 + h_1 s_7] = \frac{\sqrt{2}}{2}[3, 2, -3, 2]$$

and

$$G\mathbf{s} = [g_0 s_0 + g_1 s_1, g_0 s_2 + g_1 s_3, g_0 s_4 + g_1 s_5, g_0 s_6 + g_1 s_7] = \frac{\sqrt{2}}{2}[-1, 4, 5, -6].$$

In general, apply these operators, using the Haar wavelets, to the signal $\mathbf{s} = [s_0, s_1, \ldots, s_{m-1}]$ of length 2^n to obtain

$$(H\mathbf{s})_0 = h_0 s_0 + h_1 s_1 = \frac{\sqrt{2}}{2}(s_0 + s_1)$$

$$(H\mathbf{s})_1 = h_0 s_2 + h_1 s_3 = \frac{\sqrt{2}}{2}(s_2 + s_3)$$

$$\vdots$$

$$(H\mathbf{s})_{2^n-1} = h_0 s_{2^n-2} + h_1 s_{2^n-1} = \frac{\sqrt{2}}{2}(s_{m-2} + s_{m-1})$$

and

$$(G\mathbf{s})_0 = g_0 s_0 + g_1 s_1 = \frac{\sqrt{2}}{2}(s_0 - s_1)$$

$$(G\mathbf{s})_1 = g_0 s_2 + g_1 s_3 = \frac{\sqrt{2}}{2}(s_2 - s_3)$$

$$\vdots$$

$$(G\mathbf{s})_{2^n-1} = g_0 s_{2^n-2} + g_1 s_{2^n-1} = \frac{\sqrt{2}}{2}(s_{m-2} - s_{m-1}).$$

This is, up to a scalar, exactly the averaging and differencing previously seen in processing with Haar wavelets. The only difference is that $\sqrt{2}$ appears due to the fact that we are using normalized wavelets.

It is important to note that, in the case of the Haar wavelets, when the operator H is applied to $\mathbf{s}$ a new signal

$$\mathbf{s}^1 = [(H\mathbf{s})_0, (H\mathbf{s})_1, \ldots, (H\mathbf{s})_{2^n-1}] = \frac{\sqrt{2}}{2}[s_0 + s_1, s_2 + s_3, \ldots, s_{m-2} + s_{m-1}]$$

of length 2^{n-1} results. This new signal is half the length of the original and its entries are the coefficients of the projection of f onto V_{n-1}. When we apply G to the signal $\mathbf{s}$ we get a new signal

$$\mathbf{d}^1 = [(G\mathbf{s})_0, (G\mathbf{s})_1, \ldots, (G\mathbf{s})_{2^n-1}] = \frac{\sqrt{2}}{2}[s_0 - s_1, s_2 - s_3, \ldots, s_{m-2} - s_{m-1}]$$

whose components are the coefficients of f when projected onto $V_{n-1}^\perp$. In this way we can view H and G as projections onto the subspaces V_{n-1} and $V_{n-1}^\perp$. When the signals are concatenated ($[\mathbf{s}^1, \mathbf{d}^1]$), the function f has been decomposed as an element of $V_{n-1} \oplus V_{n-1}^\perp$.

To continue processing as in section 2.4, apply the operators H and G to the new signal $\mathbf{s}^1$. Then $\mathbf{s}^2 = H\mathbf{s}^1 = H^2\mathbf{s}$ is an element of V_{n-2} and $\mathbf{d}^2 = G\mathbf{s}^1 = G(H\mathbf{s})$ is in $V_{n-2}^\perp$. Continue processing in this manner until signals of length 1 are obtained; then string together all the processed pieces.

Returning to our example of $\mathbf{s} = [1, 2, 3, -1, 1, -4, -2, 4]$, recall that

$$\mathbf{s}^1 = H\mathbf{s} = \frac{\sqrt{2}}{2}[3, 2, -3, 2] \quad \text{and} \quad \mathbf{d}^1 = G\mathbf{s} = \frac{\sqrt{2}}{2}[-1, 4, 5, -6].$$

The next step in the processing yields

$$\mathbf{s}^2 = H\mathbf{s}^1 = \frac{\sqrt{2}}{2}\left(\frac{\sqrt{2}}{2}[5, -1]\right) = \frac{1}{2}[5, -1] \quad \text{and}$$

$$\mathbf{d}^2 = G\mathbf{s}^1 = \frac{\sqrt{2}}{2}\left(\frac{\sqrt{2}}{2}[1, -5]\right) = \frac{1}{2}[1, -5].$$

(Note that $\mathbf{s}^2$ is not a product of signals, but rather $\mathbf{s}^2 = H^2\mathbf{s} = H(H\mathbf{s})$.)

The original signal has now been transformed into

$$[\mathbf{s}^2, \mathbf{d}^2, \mathbf{d}^1] = \frac{1}{2}[5, -1, 1, -5, -\sqrt{2}, 4\sqrt{2}, 5\sqrt{2}, -6\sqrt{2}].$$

Finally, if H and G are applied to $\mathbf{s}^2$, the result is

$$\mathbf{s}^3 = H\mathbf{s}^2 = \frac{\sqrt{2}}{2} \cdot 2 \quad \text{and} \quad \mathbf{d}^3 = G\mathbf{s}^2 = \frac{\sqrt{2}}{2} \cdot 3.$$

This gives the decomposed version of our original signal, $\mathbf{s}$, as

$$\mathbf{s}^* = [\mathbf{s}^3, \mathbf{d}^3, \mathbf{d}^2, \mathbf{d}^1] = \frac{1}{2}[2\sqrt{2}, 3\sqrt{2}, 1, -5, -\sqrt{2}, 4\sqrt{2}, 5\sqrt{2}, -6\sqrt{2}].$$

In general, the result of decomposing by the operators H and G will look like

$$\mathbf{s}^* = [\mathbf{s}^n, \mathbf{d}^n, \ldots, \mathbf{d}^2, \mathbf{d}^1] = [H^m\mathbf{s}, GH^{m-1}\mathbf{s}, \ldots, GH^2\mathbf{s}, GH\mathbf{s}, G\mathbf{s}].$$

Note that this process depends only on knowing the filter coefficients $\{h_k\}$ and $\{g_k\}$. In particular, it is not necessary to know the scaling function or the mother wavelet. All the information is contained within the dilation equation which, as was stated earlier, lies at the heart of wavelet analysis. In essence, we can use wavelets without even knowing what wavelets are!

This algorithm for processing a signal via filters is called *Mallat's pyramid algorithm* [25]. The algorithm can be represented diagramatically by

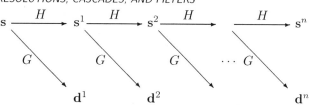

Once we know how to decompose a signal through filters, it is equally important to be able to recompose the signal. Each of the operators H and G has a so-called *dual* operator, denoted H^* and G^* respectively. These are precisely the tools needed to reverse the pyramid algorithm. The dual operators are defined by

$$(H^*\mathbf{s}^*)_k = \sum_j h_{k-2j}s_j^* \quad \text{and} \quad (G^*\mathbf{s}^*)_k = \sum_j g_{k-2j}s_j^*.$$

Note the difference in the indices from the definitions of H and G.

An example is again instructive. Let's apply these dual operators to our processed signals, $\mathbf{s}^1 = H\mathbf{s} = \frac{\sqrt{2}}{2}[3, 2, -3, 2]$ and $\mathbf{d}^1 = G\mathbf{s} = \frac{\sqrt{2}}{2}[-1, 4, 5, -6]$, to see how they "undo" the processing.

$$H^*\mathbf{s}^1 = \left[h_0 s_0^1, h_1 s_0^1, h_0 s_1^1, h_1 s_1^1, h_0 s_2^1, h_1 s_2^1, h_0 s_3^1, h_1 s_3^1\right]$$

$$= \frac{\sqrt{2}}{2}\left(\frac{\sqrt{2}}{2}[3, 3, 2, 2, -3, -3, 2, 2]\right)$$

$$= \frac{1}{2}[3, 3, 2, 2, -3, -3, 2, 2]$$

and

$$G^*\mathbf{d}^1 = \left[g_0 d_0^1, g_1 d_0^1, g_0 d_1^1, g_1 d_1^1, g_0 d_2^1, g_1 d_2^1, g_0 d_3^1, g_1 d_3^1\right]$$

$$= \frac{\sqrt{2}}{2}\left(\frac{\sqrt{2}}{2}[-1, 1, 4, -4, 5, -5, -6, 6]\right)$$

$$= \frac{1}{2}[-1, 1, 4, -4, 5, -5, -6, 6].$$

Adding the two signals yields

$$H^*\mathbf{s}^1 + G^*\mathbf{d}^1 = [1, 2, 3, -1, 1, -4, -2, 4] = \mathbf{s}.$$

Thus we have reconstructed the original signal from the first step of the processed signal. Note that H^* and G^* each produce signals that are twice the length of the input.

To see how the reconstruction process works in its entirety, consider the processed signal

$$\mathbf{s}^* = [\mathbf{s}^3, \mathbf{d}^3, \mathbf{d}^2, \mathbf{d}^1] = \left[\sqrt{2}, \frac{3\sqrt{2}}{2}, \frac{1}{2}, -\frac{5}{2}, -\frac{\sqrt{2}}{2}, 2\sqrt{2}, \frac{5\sqrt{2}}{2}, -3\sqrt{2}\right].$$

First, reconstruct $\mathbf{s}^2$ from the pair $\left[\mathbf{s}^3, \mathbf{d}^3\right]$.

$$\left[\mathbf{s}^3, \mathbf{d}^3, \mathbf{d}^2, \mathbf{d}^1\right] \rightarrow \left[H^*\mathbf{s}^3 + G^*\mathbf{d}^3, \mathbf{d}^2, \mathbf{d}^1\right]$$

$$= \left[\frac{\sqrt{2}}{2}[\sqrt{2}, \sqrt{2}] + \frac{\sqrt{2}}{2}[(\frac{3\sqrt{2}}{2}), -(\frac{3\sqrt{2}}{2})], \frac{1}{2}, -\frac{5}{2}, -\frac{\sqrt{2}}{2}, 2\sqrt{2}, \frac{5\sqrt{2}}{2}, -3\sqrt{2}\right]$$

$$= \left[\frac{5}{2}, -\frac{1}{2}, \frac{1}{2}, -\frac{5}{2}, \frac{1}{2}, -\frac{5}{2}, -\frac{\sqrt{2}}{2}, 2\sqrt{2}, \frac{5\sqrt{2}}{2}, -3\sqrt{2}\right]$$

$$= \left[\mathbf{s}^2, \mathbf{d}^2, \mathbf{d}^1\right].$$

Next, reconstruct $\mathbf{s}^1$ from the pair $\left[\mathbf{s}^2, \mathbf{d}^2\right]$.

$$\left[\mathbf{s}^2, \mathbf{d}^2, \mathbf{d}^1\right] \rightarrow \left[H^*\mathbf{s}^2 + G^*\mathbf{d}^2, \mathbf{d}^1\right]$$

$$= \left[\frac{\sqrt{2}}{2}[\frac{5}{2}, \frac{5}{2}, -\frac{1}{2}, -\frac{1}{2}] + \frac{\sqrt{2}}{2}[\frac{1}{2}, -\frac{1}{2}, -\frac{5}{2}, \frac{5}{2}], -\frac{\sqrt{2}}{2}, 2\sqrt{2}, \frac{5\sqrt{2}}{2}, -3\sqrt{2}\right]$$

$$= \left[\frac{3\sqrt{2}}{2}, \sqrt{2}, -\frac{3\sqrt{2}}{2}, \sqrt{2}, -\frac{\sqrt{2}}{2}, 2\sqrt{2}, \frac{5\sqrt{2}}{2}, -3\sqrt{2}\right]$$

$$= \left[\mathbf{s}^1, \mathbf{d}^1\right].$$

Finally, recall that earlier we saw how the original signal $\mathbf{s}$ is reconstructed from $\left[\mathbf{s}^1, \mathbf{d}^1\right]$. Note that the reconstruction process involves applying the dual operators H^* and G^* successively until we get back to the original signal.

Problems

60. If $\mathbf{s} = [s_0, s_1, \ldots, s_{m-1}]$ is a signal of length 2^n for some n, show that, for the Haar wavelets

$$H^*(H\mathbf{s}) = \frac{1}{2}\left[(s_0 + s_1), (s_0 + s_1), \ldots, (s_{m-2} + s_{m-1}), (s_{m-2} + s_{m-1})\right]$$

and

$$G^*(G\mathbf{s}) = \frac{1}{2}[(s_0 - s_1), (-s_0 + s_1), \ldots,$$

$$(s_{m-2} - s_{m-1}), (-s_{m-2} + s_{m-1})].$$

61. Show that $H^*(H\mathbf{s}) + G^*(G\mathbf{s}) = \mathbf{s}$ for any signal $\mathbf{s}$ of length 2^n using the Haar wavelets.

62. Write a paragraph explaining figure 2.7 on page 40 in chapter 2 using the idea of filters presented in this section.

Problem 61 shows that if we add $H^*(H\mathbf{s})$ to $G^*(G\mathbf{s})$ we get back our original signal. This is the last step in the reconstruction process which uses the same filter coefficients and can be represented diagramatically by

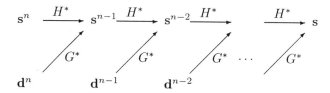

We must be careful when applying the processing and decomposing algorithms. Orthogonality plays a critical role in the projections onto the subspaces V_n and $V_n^\perp$, so these algorithms won't perform in the same way if the wavelets used are not orthogonal. For example, we won't get the same kind of results if we try this with the hat wavelets. In addition, the operators defined by (3.19) and (3.20) are designed to process signals of infinite length. So even with orthogonal wavelets there may be incorrect recomposition of the components at either end of a finite signal if the number of nonzero refinement coefficients used is different than two. This is called a *boundary problem*. One method for dealing with boundary problems, the *periodic* method, is discussed in problems 64 and 65.

=================== **Problems** ===================

63. Find the filters $\{h_k\}$ and $\{g_k\}$ for the D_4 wavelets.

64. (a) Verify that $H^*(H\mathbf{s}) + G^*(G\mathbf{s})$ returns the middle four components for a signal of length 8 using the D_4 wavelets.

    (b) One way of dealing with these boundary problems is to extend the signal beyond its boundaries, making the signal periodic. In other words, instead of working with a signal $\mathbf{s}$ of finite length, consider the signal $[\ldots, \mathbf{s}, \mathbf{s}, \mathbf{s}, \ldots]$ of infinite length. Here, let $\mathbf{s}$ be a signal of length 8. Repeat (a) with the signal $[\mathbf{s}, \mathbf{s}, \mathbf{s}]$. Show that $H^*(H\mathbf{s}) + G^*(G\mathbf{s})$ returns $\mathbf{s}$ in the middle 8 components, using the D_4 wavelets. This approach to dealing with boundary problems is called the *periodic* method. One problem with this approach is that we have extended the length of our signal.

65. (a) To define filters to operate on a finite signal without extending the signal, we use the shift operator S on a signal $\mathbf{s} = [s_0, s_1, \ldots, s_m]$. Define S by

$$S(\mathbf{s}) = [s_1, s_2, \ldots, s_n, s_0].$$

Pick a signal, $\mathbf{s}$, of length 8 and find $S^k(\mathbf{s})$ for each integer k. Note that this will involve determining the inverse, S^{-1} of the shift operator.

(b) To examine periodicity, consider the case of the D_4 wavelets. The operators H and G are defined by (3.19) and (3.20). Now define periodic operators, H_p and G_p, on a signal, s, in the following manner:

$$(H_p \mathbf{s})_k = \sum_j h_j \left(S^{2k}(\mathbf{s}) \right)_j \quad \text{and} \quad (G_p \mathbf{s})_k = \sum_j g_{j-2} \left(S^{2k-2}(\mathbf{s}) \right)_j.$$

Show that, in the case of the D_4 wavelets, these operators H_p and G_p act on a signal of length 8 exactly as the filters H and G acted on a signal when we extended it in both directions as in 64(b).

(c) The dual operators of H_p and G_p can also be defined for the D_4 wavelets, but they are more complicated. Define

$$\left(H_p^* \mathbf{s}^* \right)_k = \begin{cases} \sum_j h_{2-2j} \left(S^{\frac{k}{2}-1}(\mathbf{s}^*) \right)_j, & \text{if } k \text{ is even} \\ \sum_j h_{5-2j} \left(S^{\frac{k+1}{2}-3}(\mathbf{s}^*) \right)_j, & \text{if } k \text{ is odd} \end{cases}$$

and

$$\left(G_p^* \mathbf{s}^* \right)_k = \begin{cases} \sum_j g_{2-2j} \left(S^{\frac{k}{2}-1}(\mathbf{s}^*) \right)_j, & \text{if } k \text{ is even} \\ \sum_j g_{5-2j} \left(S^{\frac{k+1}{2}-3}(\mathbf{s}^*) \right)_j, & \text{if } k \text{ is odd.} \end{cases}$$

Remember, when H_p^* and G_p^* are used to recompose our original signal, the signal $\mathbf{s}^*$ in these formulas will be half the length of the original signal. Show that, for the D_4 wavelets,

$$H_p^*(H_p \mathbf{s}) + G_p^*(G_p \mathbf{s}) = \mathbf{s}$$

for any signal of length 8. For more details on this method or other schemes for dealing with boundary problems, see chapter 8 of [28].

3.8 MORE PROBLEMS OF THE DIGITAL AGE: COMPACT DISCS

There are two ways of representing information: analog and digital. *Analog* refers to information being presented continuously, while *digital* refers to data defined or sampled in steps. For example, in section 1.6, the graph of $f(t) = \sin(20t)(\ln(t))^2$ (analog) was compared to an approximation obtained by sampling at 32 evenly spaced points (digital). The advantage of analog information is its ability to fully represent a continuous stream of information, whereas digital data is less affected by unwanted interference or noise. When it is copied, digital information can be reproduced exactly, whereas analog information is always degraded.

Today, there is a need to be able to efficiently store sound, such as speech and music. Like images, sound can be digitized, and this is how such information is stored on compact discs. The key mathematical tool for digitization of sound is the *Shannon wavelets*. In this section, some of the ideas that have been utilized to store music on compact discs are discussed.

At a primitive level, sound is just a wave that is transmitted through a medium such as air. A simple sound, or pure tone, has the form of a sine wave. These waves are periodic — they repeat themselves. The time it takes for a wave to repeat itself is called a *cycle*.

Some waves have short cycles and others long ones, and the length of a wave's cycle is related to its *pitch*, or *frequency*. In general, sounds with short cycles have high frequency while long cycles correspond to lower frequencies. The frequency of a wave is often measured in units called Hertz (Hz), with 1 Hz meaning one cycle per second. High-pitched sounds, such as a police whistle or piccolo, have a high frequency with thousands of Hertz; low-pitched sounds, such as far-away thunder or a tuba, have a low frequency with only a few cycles per second. We hear vibrations between 20 and 20,000 Hz as sound.

Another attribute of sound is *loudness*, which is determined by the amplitude of the sound wave. Loud sounds correspond to waves with large amplitude; soft sounds correspond to waves with small amplitude. As a wave, sound can be thought of as a function s of time t, where $s(t)$ measures the *intensity* of the sound, which is the displacement of the wave from the t-axis.

Sound is a continuous phenomena. However, just as we did with the example in section 1.6, a continuous sound can be sampled and converted to a digital signal. The ideas that were discussed in chapter 1 — sampling, transforming using wavelets, and quantization — can be applied equally well to sound.

===================== **Problems** =====================

66. Suppose we want to work with trigonometric functions with frequency less than 10. Which of the following could we use? Why?

    (a) $\sin(12\pi t)$ (b) $\cos(3t)$

    (c) $\cos(43\pi t)$ (d) $\sin(70t)$

Now let's sample a sound of frequency m cycles per second (m a power of 2), where sampling at j points produces a digital signal with j components. (To maintain consistency of the digital sound throughout the signal, it is important to sample uniformly from each cycle.) First, choose a wavelet space V_n for some fixed value of n. If $2^n = m$, then the sampling can be uniform across cycles; if $2^n > m$, then each component of the signal can be divided into 2 or more subcomponents with the same intensity until the signal contains 2^n components. However, if $2^n < m$, then the sample would contain more information from some cycles than others and the signal would not represent the sound faithfully. (This is like trying to approximate a function that is defined on "sixteenths" with the Haar wavelets that are defined on "eighths.") So, in order to maintain the consistency of the signal, it is necessary to have $2^n \geq m$. The consequence is that if we start with a signal, we can choose whatever sufficiently large n we want in order to sample uniformly. However, if we fix n, then we need to limit the frequency (number of cycles per second) of our sound.

It is possible to use ideas from Fourier analysis to develop a general idea of limiting frequencies. As not all sounds are pure, they cannot be modeled by trigonometric functions, whose frequency is easily calculated (see problem 66). Instead, an analog signal $s(t)$ is called *band-limited by frequency m* if

$$\int_{-\infty}^{\infty} \cos(\omega t) s(t) \ dt = 0 \quad \text{and} \quad \int_{-\infty}^{\infty} \sin(\omega t) s(t) \ dt = 0$$

when $|\omega| > 2\pi m$.

Equivalently, s is band-limited by frequency m if

$$\hat{s}(\omega) = \int_{-\infty}^{\infty} e^{-i\omega t} s(t) \ dt = 0$$

when $|\omega| > 2\pi m$. The function $\hat{s}$ is called the *Fourier transform* of s. This is a complex-valued function, and many proofs of theorems about wavelets make use of it. Another way to say that s is band-limited is to say that its Fourier transform has compact support.

=================== **Problems** ===================

67. Prove that $f(t) = \sin(7t) + \cos(5t)$ is band-limited, and determine the limit frequency m.

68. Prove that the Haar scaling function ϕ_h is not band-limited by any frequency m.

69. It can be proven that the Fourier transform of the Shannon scaling function ϕ_s defined by $\phi_s(t) = \frac{\sin(\pi t)}{\pi t}$, is a complex multiple of $\phi_h(t + \frac{1}{2})$. Based on this fact, what can be concluded about the Shannon scaling function?

===

The Shannon Sampling Theorem states that if $s(t)$ is band-limited by frequency m, then

$$s(t) = \sum_{-\infty}^{\infty} s\left(\frac{k}{2m}\right) \frac{\sin(\pi(2mt - k))}{\pi(2mt - k)}. \tag{3.21}$$

(For a proof of this theorem[7], see [21].) Letting $m = 2^n$, (3.21) can be rewritten as

$$s(t) = \sum_{-\infty}^{\infty} s\left(\frac{k}{2^{n+1}}\right) \phi_s(2^{n+1}t - k). \tag{3.22}$$

This sampling theorem is very important. It states that for a band-limited function s, there is a frequency s_c, so that s is completely specified by its sampled values on any sampling interval of length less than $\frac{1}{2s_c}$. The frequency s_c is known as the *Nyquist frequency*. No information is lost if a signal is sampled at the Nyquist frequency, and no additional information is gained by sampling faster than this rate.

========================= **Problems** =========================

70. Explain why this theorem is called a *sampling* theorem.

71. If $s(t)$ is band-limited by frequency 1, then which values of $s(t)$ do we need to know in order to reconstruct the whole function $s(t)$? Which Shannon wavelet spaces would $s(t)$ belong to?

72. Determine the refinement coefficients for the dilation equation for the Shannon scaling function.

73. Substitute the Haar scaling function as s in (3.22), with $n = 1$. What does your result tell you about the Haar scaling function?

The next two problems suggest where wavelet analysis arises in this situation. If the sound signal $s(t)$ is band-limited by 2^{n-1}, then it can be written as a linear combination of $\{\phi_s(2^n t - k)\}$, the wavelet sons, which is sufficient for the scope of this text.

========================= **Problems** =========================

74. Prove this theorem: If s is band-limited by 2^{n-1}, then $s \in V_n$, where V_n is the nth Shannon wavelet space.

75. Prove that $s(t) = \sin(4\pi t)$ is band-limited by frequency 2. Then determine the coefficients that make s a linear combination of the appropriate wavelets.

---

[7]This result was proved by Claude Shannon in the 1940's, long before the word *wavelet* was coined.

The sampling rate of compact disc players is based on the mathematics described in this final section. The highest audible frequency is approximately 18,000 cycles per second (which is approximately $2^{14.1}$), so we can assume that any sounds that are digitized and stored on a compact disc are band-limited by, say, 2^{14}. So, for nearly all audible sounds, $s(t) \in V_{15}$, and, according to Shannon's Sampling Theorem, it follows that

$$s(t) = \sum_{-\infty}^{\infty} s\left(\frac{k}{2^{15}}\right) \phi_s(2^{15}t - k).$$

Accordingly, if we sample the sound every 2^{15} seconds, then we ought to be able to reproduce the sound exactly. In other words, a sampling frequency of 32768 cycles per second should be sufficient. In fact, compact disc players sample at 44100 cycles per second in order to cover all audible sounds, as well as some that are not audible.

There is also a quantization and coding process that takes place in the design of compact discs, similar to what was discussed in chapter 1. The function values of $s(t)$ that are sampled are converted to sixteen-bit numbers through a mapping from the range of $s(t)$ to the integers from 0 to 66535. This is done through a method called pulse code modulation (PCM), which was invented in the 1940's. For more information on PCM, see [29].

=========================== **Problems** ===========================

76. Advanced books on wavelets make considerable use of the Fourier transform. For instance, in [10], we read,

$$\hat{\phi}(\xi) = m_0(\xi/2)\hat{\phi}(\xi/2), \quad \text{where} \quad m_0(\xi) = \frac{1}{\sqrt{2}} \sum_k h_k e^{-ik\xi}.$$

Apply the Fourier transform to (3.8) in order to derive this equation from Daubechies' book.

$$\mathcal{4}$$

Sample Projects

4.1 INTRODUCTION: OVERVIEW OF PROJECTS

The projects in this chapter are for use by the reader or in a classroom.

The project in section 4.2, *Image Processing and Compression*, can be completed by students in a first semester linear algebra class using material from chapter 1. The project involves creating, processing, and compressing an image using entropy coding. A *Maple* file to generate the 16-by-16 matrix A_4 needed for the processing is included in appendix B. Students who have studied inner product spaces, orthogonal decompositions, and chapter 2 could write code to process images in this way, or can use the inverse M_{16} of A_4. This project uses two programs: *Pixel Images* for constructing and viewing images and *Maple* for processing the data.

The project in section 4.3, *A Wavelet-Based Search Engine*, also utilizes material from chapter 1, along with *Pixel Images*. In this project, wavelets are used in order to define a "distance" between images, which leads to the development of a search engine for images.

The rest of the projects are based on material in chapter 3. *B-Splines*, in section 4.4, further explores the Battle-Lemarié wavelets as they are studied in a course in numerical analysis. *Processing with the D_4 Wavelets* investigates how to use Daubechies wavelets to process images, extending the ideas of the first project. Finally, in *Daubechies Wavelets with Six Refinement Coefficients*, the reader is asked to create the D_6 scaling function in a fashion similar to how the D_4 scaling function was created in section 3.6.

4.2   LINEAR ALGEBRA PROJECT: IMAGE PROCESSING AND COMPRESSION

If we want to digitally store an image, like a picture or a fingerprint, we could partition the image vertically and horizontally and record the color or shade at each grid entry. The grid entries are considered as pixels. This gives us a matrix X of colors or gray-scale, indexed by the horizontal and vertical positions of the pixels. Each row or column can be treated as a piecewise constant function. Then we can apply a 16-by-16 wavelet conversion matrix (M_{16} or A_4) to X to convert the columns to wavelet coefficients. After that, we can choose a tolerance and eliminate all entries below that tolerance to obtain a list of vectors that we then compress, using entropy coding (table 1.1). Finally, we can convert the compressed data back to pixel shades by undoing the previous process. In this project you will apply these ideas to an image that you will construct.

Part 1: Creating an Image

Use the program *Pixel Images* (this program can be downloaded from our web site www.gvsu.edu/mathstat/wavelets.htm) to produce a gray-scale image on a 16-by-16 grid. An example is given in figure 4.1. Save this image to the file *image.txt*.

Part 2: Processing the Data

Process the data using the appropriate method (matrices A_4 and M_{16}, or inner products). Save the processed data to a file named *data1.txt*. (A *Maple* routine to read from the data file *data1.txt* is included in section B.2 in appendix B.)

Take the processed matrix from the file *data1.txt* and convert it to one long vector by concatenating the rows. Apply thresholding (hard thresholding works well) to the data to obtain a new collection of data. Save this data to a file *data2.txt*. (A *Maple* routine for writing data to a file is included in appendix B.) Use entropy coding (section 1.7) to convert to a compressed string. Save this as *compress.txt*. How much compression do you realize? Be sure to show all steps in the process.

Part 3: Reconstructing the Image

Here you must
(1) decompress the data from *compress.txt* into 256 wavelet coefficients,
(2) recompose the sixteen vectors from the 256 wavelet coefficients, and
(3) recreate the picture using *Pixel Images*.

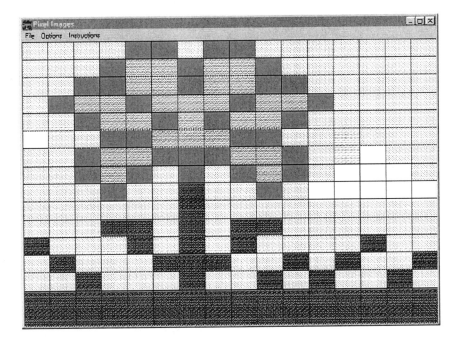

Fig. 4.1 Flower image created with *Pixel Images* by students Flora Gonzalez, Jennifer Kortjohn, and Linda Lowden from MTH 327, summer 1997 at Grand Valley State University.

Turn in the files *image.txt, data1.txt, data2.txt,* and *compress.txt* along with a picture of your original image and the image reconstructed from your compressed data.

Take care in the construction of your original image; the image reconstructed from your compressed data should strongly resemble the original image.

A report is required to accompany the files you submit. In your report you should include a discussion of all processes used in the project. Take special care when describing how you chose your thresholding value.

Part 4: Two Sided Processing (Optional)

In the previous three parts of this project we converted only the **columns** of an image matrix X to wavelet coefficients. To introduce more zeros in our matrix, we might also convert each **row** of our processed wavelet matrix $Y = AX$ to obtain their wavelet coefficients.

Explain how you could use the same matrix A to find the wavelet coefficients of the rows of Y. How would you store this final matrix?

Once you've converted both the rows and columns of your original image matrix X, how would you reconvert to obtain the original image?

Use your example from part 1 to reduce both rows and columns. Apply thresholding to this processed matrix and repeat part 3. Does this introduce significantly more zeros in the process? If not, can you explain why? Do you think this "double" processing is worthwhile? Explain.

Turn in the same three types of files as described in part 3. Label these files however you like but be sure to explicitly identify which file is which.

4.3   A WAVELET-BASED SEARCH ENGINE

Search engines on the Internet are very useful for finding certain words on World Wide Web pages. However, search engines are not nearly as efficient at locating images. In this project we will examine a way to use wavelets to design a search engine for images. The idea is motivated by the work of Wang, Wiederhold, Firchein, and Wei [43], and the writings of Stollnitz, DeRose, and Salesin [37].

For each image we can define a smaller *feature vector*, which contains much of the information from the original image. This is where wavelets come in, because wavelet coefficients can be used for that information. If we define a distance between these vectors, then the distance between images that are similar to each other will be relatively small.

================= **Problems** =================

1. Using *Pixel Images*, create a 16-by-16 grayscale image.

==

Now, we will describe how to create the feature vector $\mathbf{v} = [v_i]$, which will be in $\mathbb{R}^{80}$. The creation of this vector will require a number of transformations. First, we will use Haar wavelets (from V_4) to process the image (creating an two-dimensional image box), as described in section 2.6. This will yield a new 16-by-16 matrix $M = \{m_{i,j}\}$ of wavelet coefficients.

================= **Problems** =================

2. Explain why the entries in the upper left 8-by-8 submatrix come from averaging.

3. Why would these entries be a good proxy for the original image?

==

Apply quantile thresholding to this 8-by-8 submatrix, leaving the 48 largest (in absolute value) coefficients. Label this new 8-by-8 matrix $C = \{c_{i,j}\}$. These entries of C will be the first 64 entries in the feature vector, using the following rule:

$$\mathbf{v} = [c_{1,1}, c_{1,2}, \ldots, c_{1,8}, c_{2,1}, \ldots, c_{2,8}, \ldots, c_{8,1}, \ldots, c_{8,8}, v_{65}, v_{66}, \ldots, v_{80}]$$

For the last sixteen entries, we are going to use the diagonals of two of the submatrices of M, in particular the upper-right 8-by-8 submatrix, and the lower-left 8-by-8 submatrix. (These are the wavelet coefficients created by using a high-pass filter and a low-pass filter once.) For instance,

$$v_{65} = m_{1,9}, v_{66} = m_{2,10}, \ldots, v_{72} = m_{8,16},$$

while

$$v_{73} = m_{9,1}, v_{74} = m_{10,2}, \ldots, v_{80} = m_{16,8}.$$

By taking a sample of these wavelet coefficients, it is reasonable to conclude that if two images have nearly identical feature vectors, then the images should be quite similar.

Problems

4. Create the feature vector for your image from Problem 1.

How will we compare feature vectors from different images? We will use a weighted distance between vectors. In general, for two vectors $\mathbf{u}$ and $\mathbf{v}$, this would be defined by:

$$d(\mathbf{u}, \mathbf{v}) = \sum_{i=1}^{64} w_1 \mid u_i - v_i \mid + \sum_{i=65}^{80} w_2 \mid u_i - v_i \mid$$

where w_1 and w_2 are weights that can be chosen arbitrarily. We are going to work under the reasonable assumption that images that are similar will have similar submatrices C, and use the other coefficients as "tie-breakers." So, it will be wise to use a value of w_1 which is significantly larger than the value of w_2. For instance, we can use $w_1 = 1$ and $w_2 = \frac{1}{4}$.

Problems

5. Experiment by creating several images, some which are similar to others, and some which are quite different. Then compute their feature vectors and the distances between each pair of vectors. (For instance, four images will lead to six different distance calculations.) Report on your results.

6. This project is most interesting when several people participate, such as in a class. Compute the distance between your feature vector and several others. Discuss whether the vectors which are "close" lead to images which are "close."

7. Describe how a search engine could be designed for images.

4.4   B-SPLINES

Interpolation is the problem of finding a function that fits given data. In particular, suppose we have the values of a function f at $n+1$ data points $t_0, t_1, t_2, \ldots, t_n$. (These points are also called *nodes*.) What sort of function S can be created so that

$$S(t_0) = f(t_0), S(t_1) = f(t_1), \ldots, S(t_n) = f(t_n)?$$

There are many possible answers to this question, one of which is called a *bell-shaped spline*, or *B-spline*.

The idea behind a spline is to define a polynomial on each interval $[t_0, t_1]$, $[t_1, t_2]$, etc., so that these polynomials share the values of f at the nodes, and other conditions hold so that a certain number of derivatives of the spline exist. An example of a spline is the quadratic Battle-Lemarié scaling function, introduced in chapter 3. This spline interpolates the four function values

$$f(0) = 0, f(1) = \frac{1}{2}, f(2) = \frac{1}{2}, f(3) = 0,$$

and its derivative is continuous. Notice that this function is "bell-shaped".

=================== **Problems** ===================

1. Verify that the derivative of the quadratic Battle-Lemarié scaling function is continuous, but the second derivative is not.

==

All splines satisfy three types of conditions: *interpolation, regularity,* and *boundary*. In this section, we will focus on the specific conditions required for a cubic *B*-spline. Details about other splines can be found in various numerical analysis books [5].

A cubic *B*-spline S uses five equally-spaced nodes: t_0, t_1, t_2, t_3, t_4, but it only uses the value of the function f at t_0, t_2, and t_4. It is defined as a cubic polynomial on each of four subintervals $[t_0, t_1], [t_1, t_2], [t_2, t_3]$, and $[t_3, t_4]$. So, to begin with, we have 16 unknown constants:

$$S(t) = \begin{cases} a_0 + a_1 t + a_2 t^2 + a_3 t^3, & t_0 \le t \le t_1 \\ a_4 + a_5 t + a_6 t^2 + a_7 t^3, & t_1 \le t \le t_2 \\ a_8 + a_9 t + a_{10} t^2 + a_{11} t^3, & t_2 \le t \le t_3 \\ a_{12} + a_{13} t + a_{14} t^2 + a_{15} t^3, & t_3 \le t \le t_4. \end{cases}$$

To determine the values of these constants requires 16 equations. All of the equations turn out to be linear, so this problem can be solved using matrices.

The interpolation condition requires that S share the function values of f at the three nodes just mentioned. As a result, the following four linear equations must be satisfied:

$$a_0 + a_1 t_0 + a_2 t_0^2 + a_3 t_0^3 = f(t_0)$$
$$a_4 + a_5 t_2 + a_6 t_2^2 + a_7 t_2^3 = f(t_2) \tag{4.1}$$
$$a_8 + a_9 t_2 + a_{10} t_2^2 + a_{11} t_2^3 = f(t_2) \tag{4.2}$$
$$a_{12} + a_{13} t_4 + a_{14} t_4^2 + a_{15} t_4^3 = f(t_4).$$

The regularity condition states that S, S' and S'' must be continuous. Since S is made up of cubic polynomials, the only points where there might be trouble are the nodes t_1, t_2, and t_3. Requiring that the separate pieces of S, S' and S'' agree at these nodes leads to eight more linear equations — three each at t_1 and t_3, but only two at t_2 since (4.1) and (4.2) above guarantee that S is continuous there. For example, at t_1, there are the following three equations:

$$a_0 + a_1 t_1 + a_2 t_1^2 + a_3 t_1^3 = a_4 + a_5 t_1 + a_6 t_1^2 + a_7 t_1^3$$
$$a_1 + 2a_2 t_1 + 3a_3 t_1^2 = a_5 + 2a_6 t_1 + 3a_7 t_1^2$$
$$2a_2 + 6a_3 t_1 = 2a_6 + 6a_7 t_1.$$

Finally, the boundary conditions control what happens to the B-spline at t_0 and t_4. Basically, to make the spline "bell-shaped", we want it to flatten out at these nodes, and we can cause this to happen by making the following requirements: $S'(t_0) = S'(t_4) = 0$ and $S''(t_0) = S''(t_4) = 0$. These requirements lead to four more linear equations, for a total of sixteen.

======================= **Problems** =======================

2. Determine the cubic B-spline that interpolates the following function values: $f(3) = 1, f(6) = 8, f(9) = 1$. Graph this spline.

3. The cubic Battle-Lemarié scaling function is an example of a cubic B-spline. Its support is $0 < t < 4$, and it is known that $\phi(2) = \frac{2}{3}$. We must also have that $\phi(0) = \phi(4) = 0$ as well. Use these facts to determine the formulas for the cubic functions on the appropriate subintervals that make up this scaling function. Graph this spline.

4.5   PROCESSING WITH THE D_4 WAVELETS

In this project, we process an image constructed with the program *Pixel Images*, using the D_4 wavelets instead of the Haar wavelets. This project is similar to the Linear Algebra Project in section 4.2 and, when completed, the results should be compared to those obtained using the Haar wavelets. To undertake this project, periodic schemes as discussed in section 3.7, problems 64 and 65, must be completed.

We begin with a gray-scale matrix X indexed by the horizontal and vertical positions of the pixels. Treat each row or column as a piecewise constant function as discussed in chapter 1, then apply the D_4 wavelets to each column of the matrix to obtain wavelet coefficients.

To process using the Haar wavelets, we applied a 16-by-16 wavelet conversion matrix (M_{16} or A_4) to X to convert the columns to wavelet coefficients (see chapters 1 and 2). To apply the D_4 wavelets in a similar manner, we will need to construct the analogous matrices for these Daubachies wavelets. To construct a wavelet conversion matrix similar to the ones used to process with Haar wavelets, we will need to connect the matrix approach to the filters approach taken with the Daubachies wavelets in section 3.7. One problem we will have in applying the D_4 wavelets is that of inexact reconstruction on the boundaries of our signals. We can adapt using a periodic scheme as discussed in section 3.7, problems 64 and 65.

Part 1: Creating the Wavelet Conversion Matrix.

To produce a wavelet conversion matrix C which computes the wavelet coefficients CX from the image matrix X, using the D_4 wavelets, we need to determine how to completely process any 16-by-1 signal. As we saw in chapter 2, this is done in steps.

Step 1. Apply the periodic filters for the D_4 wavelets from problems 64 and 65 in section 3.7 to a signal, **s**, of length 16. From this, construct a 16-by-16 matrix C_1 that performs the operations of both the low pass and high pass filters associated to the D_4 wavelets.

Step 2. Find the 16-by-16 matrix C_2 that performs the operations of both the low pass and high pass filters associated to the D_4 wavelets on the first half (the result of the low pass filter) of the processed signal from Step 1. (To construct this matrix you will again need to apply the periodic filters, this time to a signal of length 8, convert the result to a matrix format, then extend to a 16-by-16 matrix that acts as the identity on the last 8 characters.)

Step 3. Find the 16-by-16 matrix C_3 that performs the operations of both the low pass and high pass filters associated to the D_4 wavelets on the first half (the result of the low pass filter) of the processed signal from Step 2.

Now combine these matrices C_1, C_2, and C_3 to construct the wavelet conversion matrix C. Use this matrix to process the image matrix, X.

To reverse the processing we need to invert the matrix C. Depending on the CAS used and the processing speed of your computer, it might be faster to invert each of C_1, C_2, and C_3 and then combine these inverses appropriately.

Part 2: Creating an Image.

Use *Pixel Images* to produce a gray-scale image on a 16-by-16 grid or use the image produced in the Haar wavelets project. Repeat the remaining parts of Project 1 using the D_4 wavelets.

Part 3 (Optional): Comparison of Haar Wavelets to D_4 Wavelets in Image Compression.

Compare the compression ratios (or the number of zeros introduced into the processed matrix after thresholding) to the quality of the recomposed image using the Haar and D_4 wavelets. Which seems to produce better compression while still maintaining the integrity of the original image? Did you expect this? Explain.

4.6   DAUBECHIES WAVELETS WITH SIX REFINEMENT COEFFICIENTS

Your goal in this project is to create and work with D_6, the scaling function of Daubechies that has six refinement coefficients. Here are the requirements for this scaling function.

1. *Compact Support.* The support of the scaling function is $0 < t < 5$.

2. *Orthogonality.* The identities from section 3.3 hold.

3. *Regularity.* Constant, linear, and quadratic functions can be written as linear combinations of elements in $\{\phi(t - k)\}$.

Part 1: Create the scaling function.

To create the scaling function, first determine the seven equations with six unknowns that arise from the conditions above. Next, solve these equations numerically using *Maple*. Then, make use of the cascade algorithm to generate D_6. You should notice that the graph of D_6 is smoother than that of D_4.

Part 2: Create the mother wavelet.

Now that the scaling function is created, you can create a graph of the mother wavelet by using the relationship between the two functions that is described in chapter 3.

Part 3: Filtering with D_6.

Problem 64 in section 3.7 describes a way to use filters based on D_4. Create the filters based on D_6 and then create a signal **s** with length appropriate to

D_6. Use the periodic method to decompose this signal with the filters. Then recompose from the wavelet coefficients.

Part 4: More Filtering with D_6 (optional).

Now apply the method of problem 65 in section 3.7 to decompose the signal. Then, recompose from the wavelet coefficients.

Appendix A
Vector Spaces and Inner Product Spaces

A.1 VECTOR SPACES

A *signal* is an ordered string of real numbers. The number of entries in the string is called the *length* of the string. We can add two signals component-wise and multiply a signal by any scalar simply by multiplying each component of the signal by that scalar. The collection of all signals of length n is denoted $\mathbb{R}^n$. This set is *closed* under addition and multiplication by scalars. By closed, we mean that if we add two signals in $\mathbb{R}^n$ or multiply a signal in $\mathbb{R}^n$ by a real number, the result is another signal in $\mathbb{R}^n$. Other familiar and useful properties are satisfied as well. For example, it does not matter in which order signals are added, addition is associative, multiplication by scalars distributes over the addition of signals, and so on. These operations of addition and multiplication by scalars make $\mathbb{R}^n$ into what is known as a *vector space*. Here is a formal definition.

Definition. A set V on which addition of elements and a multiplication by scalars (real numbers) is defined is a *vector space* and the elements of V are called *vectors* if

1. $\mathbf{u} + \mathbf{v} \in V$ for all $\mathbf{u}, \mathbf{v} \in V$ (V is closed under addition)

2. $\mathbf{u} + \mathbf{v} = \mathbf{v} + \mathbf{u}$ for all $\mathbf{u}, \mathbf{v} \in V$ (addition in V is commutative)

3. $(\mathbf{u} + \mathbf{v}) + \mathbf{w} = \mathbf{u} + (\mathbf{v} + \mathbf{w})$ for all $\mathbf{u}, \mathbf{v}, \mathbf{w} \in V$ (addition is associative in V)

4. There is an element $\mathbf{0}$ in V so that $\mathbf{0} + \mathbf{v} = \mathbf{v}$ for all $\mathbf{v} \in V$ (V contains an additive identity, an element that we label as $\mathbf{0}$)

5. For each $\mathbf{v}$ in V there is an element $-\mathbf{v}$ in V so that $\mathbf{v} + (-\mathbf{v}) = \mathbf{0}$ (each element in V has an additive inverse in V)

6. $k\mathbf{v} \in V$ for all $\mathbf{v} \in V, k \in \mathbb{R}$ (V is closed under multiplication by scalars)

7. $(kl)\mathbf{v} = k(l\mathbf{v})$ for all $\mathbf{v} \in V$, $k, l \in \mathbb{R}$ (scalars may be placed anywhere in a product)

8. $(k + l)\mathbf{v} = k\mathbf{v} + l\mathbf{v}$ for all $\mathbf{v} \in V$, $k, l \in \mathbb{R}$ (multiplication of vectors by scalars distributes over addition of scalars in V)

9. $k(\mathbf{u} + \mathbf{v}) = k\mathbf{u} + k\mathbf{v}$ for all $\mathbf{u}, \mathbf{v} \in V$, $k \in \mathbb{R}$ (multiplication by scalars distributes over addition of vectors in V)

10. $1\mathbf{v} = \mathbf{v}$ for all $\mathbf{v} \in V$.

The collection $\mathbb{R}^n$ of all signals of length n is a vector space, called n-dimensional real space. The additive identity in this space is the signal whose components are all 0. The additive inverse of a signal $[\mathbf{v}_1, \mathbf{v}_2, \ldots, \mathbf{v}_n]$ is the signal $[-\mathbf{v}_1, -\mathbf{v}_2, \ldots, -\mathbf{v}_n]$. (In chapters 1 and 2, we represent signals as column vectors. There is an obvious identification of a column vector with a row vector, which we exploit in this appendix.)

The space $\mathbb{R}^n$ is just one example of a vector space. Another is the collection of all $m \times n$ matrices with real entries under the standard addition and scalar multiplication of matrices.

An important example that is relevant to the study of wavelets is the vector space of all real valued functions of a real variable. Let

$$F = \{f : \mathbb{R} \to \mathbb{R} | f \text{ is a function}\}.$$

We define addition and multiplication by scalars on F as follows: if f and g are elements of F and k is a real number, then

$(f + g)$ is the function defined by $(f + g)(t) = f(t) + g(t)$ for all $t \in \mathbb{R}$
kf is the function defined by $(kf)(t) = k(f(t))$ for all $t \in \mathbb{R}$.

It is not hard to show that F is a vector space under these operations. The additive identity is the zero function, namely the function $z : \mathbb{R} \to \mathbb{R}$ defined by $z(t) = 0$ for all $t \in \mathbb{R}$. The additive inverse of an element $f \in F$ is the function $-f$ defined by $(-f)(t) = -(f(t))$ for all $t \in \mathbb{R}$.

Problems

1. Show that P_2, the set of all quadratic polynomials, is a vector space under the standard addition and scalar multiplication of polynomials.

2. Show that $M_{2,3}$, the collection of all 2-by-3 matrices with real entries, forms a vector space under the standard addition and scalar multiplication of matrices. Explicitly identify the additive identity. Explicitly identify the additive inverse of each element.

3. Let V_0 be the set of all functions that are constant on the interval $[0,1)$. Prove that V_0 is a vector space under the addition and scalar multiplication of functions defined in F.

A critical idea in working with vectors is whether or not a given vector can be expressed in terms of other vectors. Consider the vectors $[1,0]$ and $[0,1]$ in $\mathbb{R}^2$. What vectors may be constructed from these two vectors, using only the operations defined on the vector space? Since a vector space is closed under multiplication by scalars, all vectors of the form $a[1,0]$ and $b[0,1]$ will be in $\mathbb{R}^2$ for any real numbers a and b. Vector spaces are also closed under addition, hence each vector of the form $a[1,0] + b[0,1]$ will also be in $\mathbb{R}^2$. Such a vector is called a linear combination of $[1,0]$ and $[0,1]$.

More generally, if we have any collection of vectors $\mathbf{v}_1, \mathbf{v}_2, \ldots, \mathbf{v}_n$ in a vector space V, a vector of the form

$$\alpha_1 \mathbf{v}_1 + \alpha_2 \mathbf{v}_2 + \ldots + \alpha_n \mathbf{v}_n,$$

where $\alpha_1, \alpha_2, \ldots, \alpha_n \in \mathbb{R}$, is called a *linear combination* of $\mathbf{v}_1, \mathbf{v}_2, \ldots, \mathbf{v}_n$. Note that $[a,b] = a[1,0] + b[0,1]$ for any vector $[a,b] \in \mathbb{R}^2$. So any vector in $\mathbb{R}^2$ can be written as a linear combination of $[1,0]$ and $[0,1]$. We might say that the vectors $[1,0]$ and $[0,1]$ form a set of building blocks from which all the vectors in the space can be obtained through operations defined on the space. Such a set is called *spanning set* and we say that the set $\{[1,0], [0,1]\}$ *spans* $\mathbb{R}^2$ and write $\text{span}(\{[1,0], [0,1]\}) = \mathbb{R}^2$.

It is possible to get all the vectors in $\mathbb{R}^2$ from sets other than $\{[1,0], [0,1]\}$. For example, $[a,b] = \frac{(a+b)}{2}[1,1] + \frac{(a-b)}{2}[1,-1]$, so $\{[1,1], [1,-1]\}$ also spans $\mathbb{R}^2$. As yet another example, $[a,b] = a[1,0] + b[0,1] + 0[1,1]$, so the set $\{[1,0], [0,1], [1,1]\}$ spans $\mathbb{R}^2$ as well. However, notice that the vector $[1,1]$ is redundant. This shows that spanning sets can be larger than we need to generate the entire space. Given a spanning set that contains redundant vectors, we may remove all unnecessary vectors and arrive at a smallest set that spans the space. A set will be a *minimal spanning set* when no vector in the set can be written as a linear combination of the others. For example, in the set $\{[1,0],[0,1],[1,1]\}$, the vector [1,1] can be written as $[1,0] + [0,1]$. So [1,1] is a linear combination of [1,0] and [0,1]. A set in which no vector can be written as a linear combination of the others is said to be a *linearly independent* set.

================================ **Problems** ================================

4. Does the set $\{[2,-1],[1,-1]\}$ span $\mathbb{R}^2$? Prove your answer. If not, what set does $\{[2,-1],[1,-1]\}$ span? Is the set $\{[2,-1],[1,-1]\}$ linearly independent? Explain.

5. Does the set $\{[-1,1],[1,-1]\}$ span $\mathbb{R}^2$? Prove your answer. If not, what set does $\{[-1,1],[1,-1]\}$ span? Is the set $\{[-1,1],[1,-1]\}$ linearly independent? Explain.

A central ideal in linear algebra is the following:

Theorem. Let V be a vector space that is spanned by a finite number of vectors. Then any two minimal spanning sets for V have the same number of elements.

This theorem tells us that the number of elements in a minimal spanning set is a well defined characteristic of certain vector spaces. A vector space that is spanned by a finite number of elements is called a *finite dimensional vector space*. A minimal spanning set for a finite dimensional vector space is called a *basis* for the vector space, and the number of elements in a basis of a finite dimensional vector space is said to be the *dimension* of the vector space. A basis for a vector space is a linearly independent spanning set. It is not hard to see that $\{[1,0],[0,1]\}$ is a minimal spanning set for $\mathbb{R}^2$, so $\mathbb{R}^2$ is finite dimensional, has dimension 2, and $\{[1,0],[0,1]\}$ is a basis for $\mathbb{R}^2$. Note that $\{[1,1],[1,-1]\}$ is also a basis for $\mathbb{R}^2$.

================================ **Problems** ================================

6. What is the dimension of $\mathbb{R}^3$? What is the dimension of $\mathbb{R}^4$? What is the dimension of $\mathbb{R}^n$? Explain.

7. Find a basis for the vector space V_1 of all functions constant on the intervals $[0,\frac{1}{2})$ and $[\frac{1}{2},1)$. What is the dimension of V_1?

A.2   SUBSPACES

There are signals of every length. However, it is only the collection of signals of the same length that form a vector space (Why?). When working with signals of different lengths, we can extend the shorter signal with zeros to make it longer. For example, a signal of length 4, (such as $[1, -1, 1, 0]$), can be made into a signal of length 8 by tacking on four 0s at the end ($[1, -1, 1, 0, 0, 0, 0, 0]$). In this way we can identify $\mathbb{R}^4$ with a subset of $\mathbb{R}^8$, namely the subset $W = \{[a, b, c, d, 0, 0, 0, 0] : a, b, c, d \in \mathbb{R}\}$. Because $\mathbb{R}^4$ is a vector space, W might be one as well. It is not difficult to show that W indeed satisfies the defining properties of a vector space. Since W is a subset of a vector space ($\mathbb{R}^8$), some of the vector space properties are automatically inherited from $\mathbb{R}^8$. Specifically, properties 2, 3, 7, 8, 9, and 10, listed in the definition in section A.1, must be satisfied in W because they are satisfied in $\mathbb{R}^8$. As a result, to show W is a vector space, we only need to verify that properties 1, 4, 5, and 6 hold. This type of situation is encountered quite frequently. Since the subset W of $\mathbb{R}^8$ is a vector space in its own right, W is called a subspace of $\mathbb{R}^8$. A formal definition now follows.

Definition. Let V be a vector space. A subset W of V is a *subspace* of V if W is a vector space using the addition and scalar multiplication from V.

The discussion above hints at the following theorem.

The Subspace Theorem.   Let V be a vector space. A subset W of V is a subspace of V if

1. $\mathbf{v} + \mathbf{w} \in W$ for all $\mathbf{v}, \mathbf{w} \in W$,

2. There is an element $\mathbf{0}$ in W so that $\mathbf{0} + \mathbf{w} = \mathbf{w}$ for all $\mathbf{w} \in W$, and

3. $k\mathbf{w} \in W$ for all $\mathbf{w} \in W$, $k \in \mathbb{R}$.

Subspaces arise in many settings. Two more examples follow.

It is well-known that the sum of two continuous functions is a continuous function, any scalar multiple of a continuous function is a continuous function, and the zero function is a continuous function. The subspace theorem proves that the set of all continuous functions from $\mathbb{R}$ to $\mathbb{R}$ is a subspace of F. The same is true of the set of all differentiable functions from $\mathbb{R}$ to $\mathbb{R}$.

Next, let U be the set of all functions from $\mathbb{R}$ to $\mathbb{R}$ that are constant on the interval $[0,1]$. If we add two such functions, the sum is again constant on $[0,1]$ and if we multiply by any scalar (including 0), the product is also constant on $[0,1]$. The set U is therefore a subspace of F.

There are other examples of subspaces that are especially relevant in the study of wavelets. One is the set V_0 consisting of all functions from $\mathbb{R}$ to $\mathbb{R}$ that are constant on $[0,1)$ and 0 everywhere else. This V_0 is a subspace of U.

=================================== **Problems** ===================================

8. Let $W = \{[a, b, a + b] : a, b \in \mathbb{R}\}$. Is W a subspace of $\mathbb{R}^3$? Prove your answer. If W is a subspace of $\mathbb{R}^3$, find a basis for W.

9. Let $W = \{[a, b, ab] : a, b \in \mathbb{R}\}$. Is W a subspace of $\mathbb{R}$? Prove your answer. If W is a subspace of $\mathbb{R}^3$, find a basis for W.

10. Let V_2 be the set of all functions constant on the intervals $\left[0, \frac{1}{4}\right)$, $\left[\frac{1}{4}, \frac{1}{2}\right)$, $\left[\frac{1}{2}, \frac{3}{4}\right)$, and $\left[\frac{3}{4}, 1\right)$ and zero elsewhere. Is V_2 a subspace of F? Prove your answer. If V_2 is a subspace of F, find a basis for V_2.

A.3   INNER PRODUCT SPACES

We can think of signals in $\mathbb{R}^n$ as points or vectors in n dimensional real space. As such, we can find the distance from the origin to a signal, or the length of the vector in $\mathbb{R}^n$, via the distance formula. That is, if $\mathbf{s} = [s_1, s_2, \ldots, s_n]$ is a signal in $\mathbb{R}^n$, the distance (or the Euclidean distance) from $\mathbf{s}$ to the origin, denoted $\|\mathbf{s}\|$, is

$$\|\mathbf{s}\| = \sqrt{s_1^2 + s_2^2 + \cdots + s_n^2}.$$

This distance is related to a pseudo-product in $\mathbb{R}^n$ called the dot product. The dot product of two signals $\mathbf{r} = [r_1, r_2, \ldots, r_n]$ and $\mathbf{s} = [s_1, s_2, \ldots, s_n]$ in $\mathbb{R}^n$ is defined by

$$\mathbf{r} \cdot \mathbf{s} = s_1 r_1 + s_2 r_2 + \cdots + s_n r_n.$$

Note that $\mathbf{r} \cdot \mathbf{s}$ is related to the distance described above by

$$\|\mathbf{s}\| = \sqrt{\mathbf{s} \cdot \mathbf{s}}.$$

The dot product can be used to find angles between two vectors in $\mathbb{R}^n$. The standard formula, found in most multivariable calculus or linear algebra texts, tells us that the angle, θ, between two nonzero vectors $\mathbf{r}$ and $\mathbf{s}$ is found through the equation

$$\cos(\theta) = \frac{\mathbf{r} \cdot \mathbf{s}}{\|\mathbf{r}\| \, \|\mathbf{s}\|}.$$

=================================== **Problems** ===================================

11. Use the dot product to find the angle between the vectors $[1,0,1,1]$ and $[0,1,1,1]$ in $\mathbb{R}^4$.

12. Use the dot product to find the angle between the vectors [1,0] and [0,1] in $\mathbb{R}^2$. Should you have expected the result? Why?

The dot product satisfies certain familiar properties for any $\mathbf{r}, \mathbf{s}$, and $\mathbf{q}$ in $\mathbb{R}^n$:

1. $\mathbf{r} \cdot \mathbf{s} = \mathbf{s} \cdot \mathbf{r}$ for any $\mathbf{r}, \mathbf{s}$.

2. $(\mathbf{r} + \mathbf{s}) \cdot \mathbf{q} = \mathbf{r} \cdot \mathbf{q} + \mathbf{s} \cdot \mathbf{q}$ for any $\mathbf{q}, \mathbf{r}, \mathbf{s}$.

3. $\mathbf{s} \cdot \mathbf{s} \geq 0$ with equality if and only if $\mathbf{s} = \mathbf{0}$.

4. $k(\mathbf{r} \cdot \mathbf{s}) = (k\mathbf{r}) \cdot \mathbf{s} = \mathbf{r} \cdot (k\mathbf{s})$ for any $\mathbf{r}$ and $\mathbf{s}$, and any real number k.

The dot product is not the only "product" that satisfies these properties. For example, define $\langle \mathbf{r}, \mathbf{s} \rangle$ to be the maximum of all the quantities $|s_i - r_i|$ for i from 1 to n. It is not difficult to show that this "product" satisfies the same properties as the dot product.

It is also possible to have a "product" in vector spaces other than $\mathbb{R}^n$. Consider the vector space $F[0, 1]$ consisting of all real valued functions defined on the interval [0,1]. The set $C[0,1]$ of all continuous functions in $F[0, 1]$ is a subspace of $F[0, 1]$. For f, g in $C[0,1]$, we define

$$\langle f, g \rangle = \int_0^1 f(t)g(t)dt. \tag{A.1}$$

The basic properties of the definite integral from calculus show that this "product" also satisfies the same properties as the dot product. Any such rule is called an *inner product*. Here is a formal definition.

Definition. Let V be a vector space. An *inner product* on V is a function that assigns to each pair of vectors $\mathbf{u}, \mathbf{v}$ in V a real number, denoted $\langle \mathbf{u}, \mathbf{v} \rangle$, satisfying the following:

1. $\langle \mathbf{u}, \mathbf{v} \rangle = \langle \mathbf{v}, \mathbf{u} \rangle$ for all $\mathbf{u}, \mathbf{v} \in V$.

2. $\langle k\mathbf{u}, \mathbf{v} \rangle = k\langle \mathbf{u}, \mathbf{v} \rangle = \langle \mathbf{u}, k\mathbf{v} \rangle$ for all $\mathbf{u}, \mathbf{v} \in V$ and $k \in \mathbb{R}$.

3. $\langle \mathbf{u} + \mathbf{v}, \mathbf{w} \rangle = \langle \mathbf{u}, \mathbf{w} \rangle + \langle \mathbf{v}, \mathbf{w} \rangle$ for all $\mathbf{u}, \mathbf{v}, \mathbf{w} \in V$.

4. $\langle \mathbf{v}, \mathbf{v} \rangle \geq 0$ for all $\mathbf{v} \in V$ with equality if and only if $\mathbf{v} = \mathbf{0}$.

If V is a vector space on which an inner product, $\langle \, , \, \rangle$, is defined, the pair $(V, \langle \, , \, \rangle)$ is called an *inner product space*.

Just as with the dot product, we can define the "length" or *norm* of a vector in an inner product space and the angle between any two vectors in an inner product space. If $\mathbf{u}$ and $\mathbf{v}$ are vectors in an inner product space, then the length of the vector $\mathbf{u}$ is $\|\mathbf{u}\| = \sqrt{\langle \mathbf{u}, \mathbf{u} \rangle}$, and the angle, θ, between $\mathbf{u}$ and $\mathbf{v}$ is given by

$$\cos(\theta) = \frac{\langle \mathbf{u}, \mathbf{v} \rangle}{\|\mathbf{u}\|\|\mathbf{v}\|}.$$

For example, in the space $C[0, 1]$,

$$\|t\| = \sqrt{\int_0^1 t^2 dt} = \sqrt{\left.\frac{t^3}{3}\right|_0^1} = \sqrt{\frac{1}{3}} \quad \text{and} \quad \|t^2\| = \sqrt{\int_0^1 t^4 dt} = \sqrt{\left.\frac{t^5}{5}\right|_0^1} = \sqrt{\frac{1}{5}}.$$

A key idea in working with vectors in inner product spaces is the notion of orthogonality. Two vectors in an inner product space are *orthogonal* if the angle between the vectors is $\frac{\pi}{2}$ radians. This is a generalization of the concept of perpendicular vectors in $\mathbb{R}^n$. A consequence of the formula for the angle between two vectors is the following theorem.

Theorem. Two nonzero vectors $\mathbf{u}$ and $\mathbf{v}$ in an inner product space are orthogonal if and only if $\langle \mathbf{u}, \mathbf{v} \rangle = 0$.

=================== **Problems** ===================

13. The notion of the angle between functions in the inner product space $C[0, 1]$ differs from the idea of the angle between curves as we discuss it in calculus. When we graph functions in Euclidean space, we are concerned about orientation, or the direction in which we move as we travel along a curve. Let f and g be functions in $C[0, 1]$ defined by $f(t) = t$ and $g(t) = 1 - t$.

   (a) Draw the graphs of these functions and find the angle between them in the plane.

   (b) As vectors in $C[0, 1]$ we no longer care what the graphs of the functions look like, we simply think of them as elements of a set. Now find the "angle" between the functions using the inner product (A.1). Do your results agree?

14. (a) Which of the following pairs f, g are orthogonal in $C[0, 1]$? Explain.

   i. $f(t) = 2 + t$, $g(t) = 1 - t$
   ii. $f(t) = t^2$, $g(t) = t - \frac{3}{4}$

   (b) Define $f, g \in C[0, 1]$ by $f(t) = at + 1$ and $g(t) = bt - 1$. Using the inner product (A.1), find all values of a and b for which f and g are orthogonal. Draw graphs to illustrate.

A.4   THE ORTHOGONAL DECOMPOSITION THEOREM

The notion of orthogonality leads us to consider the notion of orthogonal subspaces. To illustrate, let $I = \{(a, 0) : a \in \mathbb{R}\}$ and $J = \{(0, b) : b \in \mathbb{R}\}$. Then I is the x-axis in the plane and J is the y-axis. It is not difficult to show that I and J are subspaces of $\mathbb{R}^2$. If we take the dot product of the vector $(a, 0)$ in I and the vector $(0, b)$ in J, we get $(a, 0) \cdot (0, b) = a0 + 0b = 0$. This shows that every nonzero vector in I is orthogonal to every nonzero vector in J. When this happens we say that I and J are *orthogonal subspaces* in $\mathbb{R}^2$.

More generally, let V be an inner product space and W a subspace of V. An important subspace of V related to W is the *orthogonal complement* of W in V. The orthogonal complement of W is the set

$$W^\perp = \{\mathbf{v} \in V : \langle \mathbf{v}, \mathbf{w} \rangle = 0 \text{ for all } \mathbf{w} \in W\}.$$

Verifying that $W^\perp$ is in fact a subspace of V is simply a matter of applying the definition of $W^\perp$ and properties of inner products.

Examples:

1. Let $V = \mathbb{R}^2$ with the dot product as its inner product and

$$W = \{[a, 0] : a \in \mathbb{R}\}.$$

   Then $W^\perp = \{[0, b] : b \in \mathbb{R}\}$.

2. Let $V = C[0, 1]$ with the integral inner product and W the set of all functions constant on $[0,1]$. Then $f \in W^\perp$ if $\int_0^1 cf(t)dt = 0$ for any constant c. This will only happen if $\int_0^1 f(t)dt = 0$. Thus

$$W^\perp = \{f \in C[0, 1] : \int_0^1 f(t)dt = 0\}.$$

   An example of a function in $W^\perp$ is $f(t) = 1 - 2t$.

Orthogonality makes certain computations in inner product spaces straightforward. For example, suppose $B = \{\mathbf{v}_1, \mathbf{v}_2, \ldots, \mathbf{v}_n\}$ is a basis for an inner product space in which distinct vectors are orthogonal. In other words, $\langle \mathbf{v}_i, \mathbf{v}_j \rangle = 0$ if $i \neq j$. Such a basis is called an *orthogonal basis*. Suppose we choose an element $\mathbf{v} \in V$. Since B is a basis for V, we can write

$$\mathbf{v} = \alpha_1 \mathbf{v}_1 + \alpha_2 \mathbf{v}_2 + \cdots + \alpha_n \mathbf{v}_n = \sum_{i=1}^{n} \alpha_i \mathbf{v}_i$$

for some real numbers $\alpha_1, \alpha_2, \ldots, \alpha_n$. Then

$$\langle \mathbf{v}, \mathbf{v}_i \rangle = \left\langle \sum_{j=1}^{n} \alpha_j \mathbf{v}_j, \mathbf{v}_i \right\rangle = \sum_{j=1}^{n} \alpha_j \langle \mathbf{v}_j, \mathbf{v}_i \rangle = \alpha_i \langle \mathbf{v}_i, \mathbf{v}_i \rangle.$$

Thus, $\alpha_i = \frac{\langle \mathbf{v}, \mathbf{v}_i \rangle}{\langle \mathbf{v}_i, \mathbf{v}_i \rangle}$ for each i and

$$\mathbf{v} = \frac{\langle \mathbf{v}, \mathbf{v}_1 \rangle}{\langle \mathbf{v}_1, \mathbf{v}_1 \rangle} \mathbf{v}_1 + \frac{\langle \mathbf{v}, \mathbf{v}_2 \rangle}{\langle \mathbf{v}_2, \mathbf{v}_2 \rangle} \mathbf{v}_2 + \cdots + \frac{\langle \mathbf{v}, \mathbf{v}_n \rangle}{\langle \mathbf{v}_n, \mathbf{v}_n \rangle} \mathbf{v}_n = \sum_{i=1}^{n} \frac{\langle \mathbf{v}, \mathbf{v}_i \rangle}{\langle \mathbf{v}_i, \mathbf{v}_i \rangle} \mathbf{v}_i \quad (A.2)$$

This tells us that to compute the coefficients of a given vector with respect to an orthogonal basis, we need only calculate $2n$ inner products, where n is the number of elements in the basis.

=================================== **Problems** ===================================

15. Let $B = \{[1,0,1],[0,1,0],[-1,0,1]\}$. Show that B is an orthogonal basis for $\mathbb{R}^3$. Find the coefficients of the vector [1,2,3] with respect to the basis B.

The formula (A.2) can be simplified under certain conditions. If each of the vectors $\mathbf{v}_i$ satisfies $\|\mathbf{v}_i\| = \sqrt{\langle \mathbf{v}_i, \mathbf{v}_i \rangle} = 1$ (or has norm equal to 1), then (A.2) becomes

$$\mathbf{v} = \langle \mathbf{v}, \mathbf{v}_1 \rangle \mathbf{v}_1 + \langle \mathbf{v}, \mathbf{v}_2 \rangle \mathbf{v}_2 + \cdots + \langle \mathbf{v}, \mathbf{v}_n \rangle \mathbf{v}_n = \sum_{i=1}^{n} \langle \mathbf{v}, \mathbf{v}_i \rangle \mathbf{v}_i.$$

An orthogonal basis that has the property that every basis vector has norm 1 is called an *orthonormal basis*. With an orthonormal basis, computations are quick and easy.

An important question arises here. Is it always possible to find an orthogonal basis for an inner product space? The answer is yes. There is a method by which an orthogonal basis for a finite dimensional inner product space can be constructed from any basis. This method is called the *Gram–Schmidt process* and can be found in most linear algebra texts.

We can take the idea of orthogonality even further. Suppose V is a finite dimensional inner product space and W is a subspace of V. Now W will have a basis and from it we can construct an orthogonal basis for W, say $B = \{\mathbf{w}_1, \mathbf{w}_2, \ldots, \mathbf{w}_k\}$. If $\mathbf{v} \in V$, then there is a vector $\mathbf{v}_B$ in W associated to $\mathbf{v}$, namely

$$\mathbf{v}_B = \frac{\langle \mathbf{v}, \mathbf{w}_1 \rangle}{\langle \mathbf{w}_1, \mathbf{w}_1 \rangle} \mathbf{w}_1 + \cdots + \frac{\langle \mathbf{v}, \mathbf{w}_k \rangle}{\langle \mathbf{w}_k, \mathbf{w}_k \rangle} \mathbf{w}_k = \sum_{i=1}^{k} \frac{\langle \mathbf{v}, \mathbf{w}_i \rangle}{\langle \mathbf{w}_i, \mathbf{w}_i \rangle} \mathbf{w}_i. \quad (A.3)$$

Of what use is this vector $\mathbf{v}_B$? Let's look at an example.

Let $\mathbf{w}_1 = [1,1,1]$, $\mathbf{w}_2 = [0,1,-1]$, and $B = \{\mathbf{w}_1, \mathbf{w}_2\}$. Then $W = \text{span}(B)$ is a subspace of $\mathbb{R}^3$. Since $\mathbf{w}_1 \cdot \mathbf{w}_2 = 0$, B is an orthogonal basis for W. Note that, since $\|\mathbf{w}_1\| = \sqrt{\mathbf{w}_1 \cdot \mathbf{w}_1} = \sqrt{3}$ and $\|\mathbf{w}_2\| = \sqrt{\mathbf{w}_2 \cdot \mathbf{w}_2} = \sqrt{2}$, B is not an orthonormal basis for W. However, the set

$$B' = \left\{ \frac{\mathbf{w}_1}{\|\mathbf{w}_1\|}, \frac{\mathbf{w}_2}{\|\mathbf{w}_2\|} \right\} = \left\{ \left[\frac{1}{\sqrt{3}}, \frac{1}{\sqrt{3}}, \frac{1}{\sqrt{3}} \right], \left[0, \frac{1}{\sqrt{2}}, -\frac{1}{\sqrt{2}} \right] \right\}$$

is an orthonormal basis for W. If $\mathbf{v} = [3, -1, 2]$, then (A.3) yields

$$\mathbf{v}_B = \frac{\mathbf{v} \cdot \mathbf{w}_1}{\|\mathbf{w}_1\|^2}\mathbf{w}_1 + \frac{\mathbf{v} \cdot \mathbf{w}_2}{\|\mathbf{w}_2\|^2}\mathbf{w}_2 = \frac{4}{3}[1, 1, 1] + \frac{-3}{2}[0, 1, -1] = \left[\frac{4}{3}, -\frac{1}{6}, \frac{17}{6} \right].$$

Since $\mathbf{v}_B$ is a linear combination of $\mathbf{w}_1$ and $\mathbf{w}_2$, it follows that $\mathbf{v}_B \in W$. If we let $\mathbf{z} = \mathbf{v} - \mathbf{v}_B$, then a straightforward calculation shows that $\langle \mathbf{z}, \mathbf{w} \rangle = 0$ for any $\mathbf{w} \in W$. This shows that $\mathbf{v}_B$ is in W while $\mathbf{v} - \mathbf{v}_B$ is in $W^\perp$. Hence we are able to decompose the vector $\mathbf{v}$ into a sum of a vector in W and a vector in $W^\perp$. A natural question to ask is, given any subspace W and vector $\mathbf{v}$, can we decompose $\mathbf{v}$ into such a sum?

Problems

16. Show that $\langle \mathbf{z}, \mathbf{w} \rangle = 0$ for any $\mathbf{w} \in W$, where $\mathbf{z}$ is defined as in the previous example.

17. Let $B = \{[0,0,1,1], [1,1,0,0], [-1,1,1,-1]\}$ and let $W = \text{span}(B)$. Show that B is an orthogonal basis for W and find $\mathbf{v}_B$ if $\mathbf{v} = [2,-1,3,1]$. Prove that $\mathbf{v} - \mathbf{v}_B$ is in $W^\perp$.

18. Let $B = \{t, 3t - 2\}$ and let $W = \text{span}(B)$. Consider W to be a subspace of the inner product space $C[0, 1]$ using the inner product (A.1). Show that B is an orthogonal basis for W and find g_B if $g(t) = t^3$. Prove that $g - g_B$ is in $W^\perp$.

The formula (A.3) will allow us to show that the decomposition of a vector as mentioned earlier always exist. The subsequent discussion will lead us to an important theorem that guarantees this result. We have seen that if $B = \{\mathbf{w}_1, \mathbf{w}_2, \ldots, \mathbf{w}_k\}$ is an orthogonal basis for W and if $\mathbf{v} \in V$, then

$$\mathbf{v}_B = \sum_{i=1}^{k} \frac{\langle \mathbf{v}, \mathbf{w}_i \rangle}{\langle \mathbf{w}_i, \mathbf{w}_i \rangle}\mathbf{w}_i \in W.$$

The vector $\mathbf{v}_B$ is called the *projection* of $\mathbf{v}$ onto W. To show that $\mathbf{v} - \mathbf{v}_B$ is a vector in $W^\perp$, we need to verify that $\langle \mathbf{v} - \mathbf{v}_B, \mathbf{w} \rangle = 0$ for all $\mathbf{w} \in W$. To do so, let $\mathbf{w} \in W$. Then

$$\mathbf{w} = \beta_1\mathbf{w}_1 + \beta_2\mathbf{w}_2 + \ldots + \beta_k\mathbf{w}_k = \sum_{j=1}^{k} \beta_j\mathbf{w}_j$$

for some real numbers $\beta_1, \beta_2, \ldots, \beta_k$. Keeping in mind that B is an orthogonal basis, we have

$$\langle \mathbf{v} - \mathbf{v}_B, \mathbf{w} \rangle = \langle \mathbf{v}, \mathbf{w} \rangle - \langle \mathbf{v}_B, \mathbf{w} \rangle$$

$$= \langle \mathbf{v}, \mathbf{w} \rangle - \left\langle \mathbf{v}_B, \sum_{j=1}^{k} \beta_j \mathbf{w}_j \right\rangle$$

$$= \langle \mathbf{v}, \mathbf{w} \rangle - \sum_{j=1}^{k} \beta_j \langle \mathbf{v}_B, \mathbf{w}_j \rangle$$

$$= \langle \mathbf{v}, \mathbf{w} \rangle - \sum_{j=1}^{k} \beta_j \left\langle \sum_{i=1}^{k} \frac{\langle \mathbf{v}, \mathbf{w}_i \rangle}{\langle \mathbf{w}_i, \mathbf{w}_i \rangle} \mathbf{w}_i, \mathbf{w}_j \right\rangle$$

$$= \langle \mathbf{v}, \mathbf{w} \rangle - \sum_{j=1}^{k} \beta_j \left\langle \frac{\langle \mathbf{v}, \mathbf{w}_j \rangle}{\langle \mathbf{w}_j, \mathbf{w}_j \rangle} \mathbf{w}_j, \mathbf{w}_j \right\rangle$$

$$= \langle \mathbf{v}, \mathbf{w} \rangle - \sum_{j=1}^{k} \beta_j \frac{\langle \mathbf{v}, \mathbf{w}_j \rangle}{\langle \mathbf{w}_j, \mathbf{w}_j \rangle} \langle \mathbf{w}_j, \mathbf{w}_j \rangle$$

$$= \langle \mathbf{v}, \mathbf{w} \rangle - \sum_{j=1}^{k} \beta_j \langle \mathbf{v}, \mathbf{w}_j \rangle$$

$$= \langle \mathbf{v}, \mathbf{w} \rangle - \langle \mathbf{v}, \sum_{j=1}^{k} \beta_j \mathbf{w}_j \rangle$$

$$= \langle \mathbf{v}, \mathbf{w} \rangle - \langle \mathbf{v}, \mathbf{w} \rangle$$

$$= 0.$$

Hence, we have shown that $\mathbf{v} - \mathbf{v}_B$ is in $W^\perp$. The vector defined by (A.3) is called the *projection of v onto* W and the difference $\mathbf{v} - \mathbf{v}_B$ is the *projection of v perpendicular* (or *orthogonal*) to W. (The difference is also called the *residual.*) The general result is known as the *Orthogonal Decomposition Theorem*, which may be stated as follows:

The Orthogonal Decomposition Theorem. If W is a finite-dimensional subspace of an inner product space V, then any $\mathbf{v} \in V$ can be written uniquely as $\mathbf{v} = \mathbf{w} + \mathbf{w}_\perp$, where $\mathbf{w} \in W$ and $\mathbf{w}_\perp \in W^\perp$.

We represent this theorem symbolically by writing $V = W \oplus W^\perp$. Note that we have shown that if $\{\mathbf{w}_1, \mathbf{w}_2, \ldots, \mathbf{w}_k\}$ is an orthogonal basis for W, then $\mathbf{w}$, as described in the Orthogonal Decomposition Theorem, is found by

$$\mathbf{w} = \frac{\langle \mathbf{v}, \mathbf{w}_1 \rangle}{\langle \mathbf{w}_1, \mathbf{w}_1 \rangle} \mathbf{w}_1 + \cdots + \frac{\langle \mathbf{v}, \mathbf{w}_k \rangle}{\langle \mathbf{w}_k, \mathbf{w}_k \rangle} \mathbf{w}_k = \sum_{i=1}^{k} \frac{\langle \mathbf{v}, \mathbf{w}_i \rangle}{\langle \mathbf{w}_i, \mathbf{w}_i \rangle} \mathbf{w}_i$$

and $\mathbf{w}_\perp = \mathbf{v} - \mathbf{w}$. The only thing we have not yet verified is the uniqueness. This issue is addressed in the exercises.

_____ **Problems** _____

19. If V is an inner product space and W is a subspace of V, what is $W \cap W^\perp$? Prove your answer.

20. Prove the uniqueness portion of the Orthogonal Decomposition Theorem.

    Hint: Use the result from problem 19.

21. Let V_1 be the subspace of the inner product space $F[0, 1]$ (using the integral inner product) consisting of all functions constant on the intervals $[0, \frac{1}{2})$ and $[\frac{1}{2}, 1)$. Let ϕ and ψ be defined by

$$\phi(t) = \begin{cases} 1, & \text{if } 0 \le t < \frac{1}{2} \\ 1, & \text{if } \frac{1}{2} \le t < 1 \end{cases} \quad \text{and} \quad \psi(t) = \begin{cases} 1, & \text{if } 0 \le t < \frac{1}{2} \\ -1, & \text{if } \frac{1}{2} \le t < 1 \end{cases}.$$

    (a) Show that $B = \{\phi, \psi\}$ is an orthogonal basis for V_1. Is B an orthonormal basis? Explain. If not, construct an orthonormal basis for V_1 from B.

    (b) Let $f \in F[0, 1]$ be defined by $f(t) = t^2 + 1$. Use the Orthogonal Decomposition Theorem to find f_{V_1} and $f_{V_1^\perp}$. Verify that

$$f = f_{V_1} + f_{V_1^\perp}.$$

Appendix B
Maple Routines

This appendix contains *Maple* routines for some of the problems and projects in the book. *Maple* commands are indicated on the lines beginning with the > symbol. All *Maple* commands end with a colon (:) or a semi-colon (;).

B.1   MATRIX GENERATOR

This loop will generate the wavelet processing matrices, $M = M_n$ and $A = A_k$, for $n = 2^k$, where k is specified in the second line of the algorithm below. During the process, matrices M.t are created that can be used to perform the various steps in the processing stages as discussed in section 2.6. These matrices are responsible for the images in figures 2.6 and 2.7. Note that $M_n^{-1} = A_k$.

```
> with(linalg):
> k:=5: n.0:=2^k: for t from 1 to k do n.t:=n.0/2^t: od:
> for t from 0 to k-1 do M.t := matrix(n.0,n.0,proc (i,j) if
> (i<=(n.t/2) and  (2*i-1)<=j and j<=(2*i)) then (1/2) elif
> (i>(n.t/2) and i<=n.t and (2*(i-n.t/2)-1)<=j
> and j<=(2*(i-n.t/2)))
> then (1/2)*(-1)^(j+1)
```

```
> elif (i>n.t and i=j) then 1 else 0 fi end:
> od:
> M:=M.0: A:=inverse(M.0):
> for q from 1 to k-1 do M := multiply(M.q,M):
> A := multiply(A,inverse(M.q)): od:
```

B.2 PROCESSING SAMPLED DATA

We will need the `linalg` and `plots` packages.

```
> with(linalg): with(plots):
```

Since we convert the data to strings of length 8 for processing, we need to input the 8-by-8 wavelet conversion matrix A8 and its inverse, A8inv. We can compute this as shown in the matrix generator routine.

Before implementing the *Maple* routine below we must first compute A8 and A8inv as discussed above.

Now define the function to be sampled using *Maple*'s function syntax.

```
> f := t -> sin(20*t)*ln(t)^2;
> plot(f(t),t=0..1);
```

Next sample the function at N (chosen here to be 32) evenly spaced points. The storage location `data` contains the function values at the N points, and `pts` is a list of the data points for plotting purposes.

```
> N := 32;
> data := [seq(evalf(f(i/N)),i=1..N)]:
> pts := [seq([i/N,data[i]],i=1..N)]:
```

To plot the data points, use the `plot` command.

```
> plot(pts,x=0..1,color=blue);
```

Save the data plot in a variable `data_plot` to compare to processed data later.

```
> data_plot:=plot(pts,x=0..1,color=blue):
```

Collect the data in groups of 8.

```
> for i from 1 to N/8 do
> data.i := matrix(8,1,(r,s) -> data[8*(i-1)+r]); od;
```

Now we convert to wavelet coefficients by multiplying by the wavelet conversion matrix A8.

```
> for i from 1 to N/8 do wave_data.i:=multiply(A8,data.i): od:
```

Maple computes exact values when possible. The next loop uses the `evalf` command to convert to decimal form.

```
> for i from 1 to N/8 do wave_data.i:=map(evalf,wave_data.i); od;
```

To process the data, set a tolerance to eliminate some data. Each data entry whose absolute value is less than the tolerance level gets set to 0.

```
> tolerance := 0.05;
> for i from 1 to N/8 do for j from 1 to 8 do if
> abs(wave_data.i[j,1])<tolerance
> then wave_data.i[j,1]:=0: fi: od: od:
```

The next loop prints the processed data after keep or kill.

```
> for i from 1 to N/8 do evalm(wave_data.i); od;
```

To deprocess the processed data apply `A8inv`.

```
> for i from 1 to N/8 do newdata.i := multiply(A8inv,wave_data.i): od:
```

The reconstructed data after thresholding is stored in `recon_data`.

```
> recon_data := [seq(seq(newdata.i[j,1],j=1..8),i=1..N/8)]:
```

To plot, the reconstructed data is saved as `new_pts`.

```
> new_pts := [seq([i/N,recon_data[i]],i=1..N)]:
```

The next commands plot the reconverted data points as `newdata_plot` and compare to the original data plot.

```
> newdata_plot:=plot(new_pts,x=0..1,color=red):
> display({data_plot,newdata_plot});
```

B.3   PROJECTIONS ONTO WAVELET SPACES

First define the father and mother wavelets.

```
> phi := t -> piecewise(0 <= t and t < 1, 1, t < 0 or 1 <= t, 0);
> psi := t -> piecewise(0 <= t and t < 1/2, 1, 1/2 <= t
> and t<= 1, -1, t < 0 or 1 < t, 0);
```

Define the inner product on $L^2[0,1]$.

```
> ip := (f,g) -> int(f*g,t=0..1);
```

Then define the general daughter wavelet.

```
> psi_dt := (t,j,k) -> psi(2^j*t-k);
```

Now define a function to project onto wavelet space.

```
> f := t -> cos(2*Pi*t);
> plot(f(t),t=0..1);
```

Specify n to project onto V_n.

```
> n := 4;
```

Next define the wavelet sons.

```
> for r from 0 to n-1 do for s from 0 to 2^r-1 do
> psi.r.s := psi_dt(t,r,s); od; od;
```

Since B_n is an orthogonal basis for V_n, to find the wavelet coefficients a.r.s simply compute inner products. Then add the various projections together. The next loop does that.

```
> proj := ip(f(t),phi(t))*phi(t):
> for r from 0 to n-1 do for s from 0 to 2^r-1 do
> a.r.s := ip(f(t),psi.r.s)/ip(psi.r.s,psi.r.s);
> proj := proj + a.r.s*psi.r.s; od; od;
```

Plot the graphs to compare.

```
> plot({f(t),proj},t=0..1);
```

B.4   THE CASCADE ALGORITHM

First enter the function to use as the starting point for the procedure. Here, the tent function is used. (Note: the piecewise command assigns a function the value of 0 on any interval on which it has not been defined.)

```
> f := t -> piecewise(-1<=t and t<0,1+t,0<=t and t<1,1-t);
```

Set this function to the base step, f.0.

```
> f.0 := f(t):
```

Enter the dilation coefficients in a vector c. In this loop the dilation coefficients for the D_4 wavelet are used.

```
> c:=[(1+sqrt(3))/4,(3+sqrt(3))/4,(3-sqrt(3))/4,(1-sqrt(3))/4];
```

Then set bounds for the sum.

```
> c_init:=0: c_fin:=3:
```

This loop performs the iterations, $f_{j+1} = \sum_{c\_init}^{c\_fin} c_k f_j(2t - k)$ a total of N times.

```
> N := 3: for j from 0 to N-1 do
> f.(j+1) := sum(c[k+1]*subs(t=2*t-k,f.j), k=c_init..c_fin):
> od:
```

To create an animation of the process, load the plots package.

```
> with(plots):
```

The following loop constructs the graphs of the iterates created through the cascade algorithm.

```
> for q from 0 to 3 do p.q:=plot(f.q,t=-1..4): od:
```

The next lines create a vector containing these plots and then displays them in sequence.

```
> L := [seq(p.s,s=0..3)]:
> display(L,insequence=true);
```

B.5 PROCESSING AN IMAGE FROM *Pixel Images*

First create a grayscale image with the *Pixel Images* program. Save this image as a file, *image.txt*. In *Maple*, load the `linalg` package.

```
> with(linalg):
```

Since we are working with a 16-by-16 grayscale image, we next need to input the 16-by-16 wavelet conversion matrix A16 and its inverse, A16inv. Compute this matrix as shown in the matrix generator routine.

The following lines import the image matrix X from *Pixel Images*. If you saved your image matrix in the file *image.txt*, enter that in fname. Otherwise, enter whatever name you gave to your image matrix. You need to include the drive designation. If you have your file saved on a disk in the a: drive, your file name will be `'a: \image.txt'`. The ' needed here is the single back quotation mark. The space between the colon and the file name seems to be necessary on some systems.

```
> fname := 'a: \image.txt';
> X := matrix(16,16,(i,j) -> 0):
```

```
> for i from 1 to 16 do for j from 1 to 16 do
> X[i,j]:=fscanf(fname,`%d`)[1]: od: od: fclose(fname):
> evalm(X);
```

The image data is processed by multiplying by **A16**. The processed data is saved in the matrix **Y**.

```
> Y := multiply(A16,X);
```

Convert the entries to decimal form to make them easier to read.

```
> Y := map(evalf,Y);
```

Now apply thresholding. Select a tolerance and eliminate those entries below the tolerance. The new data is saved in the matrix **Z**.

```
> tolerance := 1;
> Z := matrix(16,16,(i,j) -> if abs(Y[i,j])<=tolerance then 0
> else Y[i,j] fi);
```

Deprocess the data by multiplying by **A16inv**. The deprocessed data is saved in **newZ**.

```
> newZ := multiply(A16inv,Z);
```

Pixel Images requires that data be positive integers. Convert the processed data to absolute values and round to the nearest integer.

```
> newZ := map(abs,newZ);
> newZ := map(round,newZ);
```

Write the matrix **newZ** to an output file **f_out** so that the processed image can be viewed in *Pixel Images*. Again, include the drive designation. Then print this data to the named file. *Pixel Images* can then be used to compare the original picture to the processed picture.

```
> f_out := `a: \image2.txt`;
> for i from 1 to 16 do for j from 1 to 16 do
> fprintf(f_out,`%d\n`,newZ[i,j]): od: od:
> fclose(f_out):
```

Appendix C
Answers to Selected Problems

Chapter 1

2. A 3 inch by 5 inch black-and-white photograph in 8-bit grayscale at 500 dpi generates 3,750,000 bytes, or 3.75 MB of data. (The Institute for Electrical and Electronics Engineers (IEEE) has proposed the terminology "mebibyte" (MiB) to stand for 2^{20} bytes of data, since computers work in binary rather than base 10. A floppy disk actually contains 1.44 mebibyte of data. So, another acceptable answer would be 3.58 MiB. For more information about the proposed terminology, see http://physics.nist.gov/cuu/Units/binary.html.)

3. A 3 inch by 5 inch color photograph using 24-bit colors at 500 dpi generates 11,250,000 bytes, or 11.25 MB of data. (Or, 10.73 MiB.)

4. If

$$[12, 2, -5, 15]^T = x_1 \begin{bmatrix} 1 \\ 0 \\ 0 \\ 0 \end{bmatrix} + x_2 \begin{bmatrix} 0 \\ 1 \\ 0 \\ 0 \end{bmatrix} + x_3 \begin{bmatrix} 0 \\ 0 \\ 1 \\ 0 \end{bmatrix} + x_4 \begin{bmatrix} 0 \\ 0 \\ 0 \\ 1 \end{bmatrix},$$

then $x_1 = 12$, $x_2 = 2$, $x_3 = -5$, and $x_4 = 15$.

5. In this case, the coefficients x_1, x_2, x_3, and x_4 are 6, 1, 5, and -10, respectively.

12. The set V_n consists of functions that are piecewise constant on intervals of the form $\left[\frac{i}{2^n}, \frac{i+1}{2^n}\right)$ for i from 0 to $2^n - 1$. From this we can see that V_n corresponds to $\mathbb{R}^{2^n}$.

13. (a) (i) To write $[3, 7, -4, -6]^T$ as $x_1\phi(t) + x_2\psi(t) + x_3\psi_{1,0}(t) + x_4\psi_{1,1}(t)$
we need $x_1 = 0$, $x_2 = 5$, $x_3 = -2$, and $x_4 = 1$.
   (ii) In this case we obtain $x_1 = 2$, $x_2 = 27$, $x_3 = -15$, and $x_4 = 0$.

16. (a) (i) The wavelet coefficents are 1, -1, 5, 27, -2, 1, -15, and 0.
   (ii) The wavelet coefficients are $\frac{989}{8}$, $\frac{931}{8}$, 10, $\frac{11}{4}$, 0, 10, 5, and $\frac{1}{2}$.

23. The signal $\mathbf{e}$ is given by

$$\mathbf{e} = [110, 140, 181, 6, 214, 103, 7, 209, 226, 104, 136, 172, 5]^T.$$

Chapter 2

2. (a) The inner product $\langle t, \phi(t) \rangle$ is $\frac{1}{2}$.
   (b) The inner product $\langle \psi_{1,0}(t), \psi_{1,1}(t) \rangle$ is 0. Note that this shows $\psi_{1,0}$ and $\psi_{1,1}$ are orthogonal.

3. (a) The norm $\|\psi_{1,0}(t)\|$ is $\frac{\sqrt{2}}{2}$.
   (b) The norm $\|\psi_{2,1}(t)\|$ is $\frac{1}{2}$.

9. The orthogonal projection of $h(t) = t$ onto V_2 is

$$f(t) = \frac{1}{2}\phi - \frac{1}{4}\psi - \frac{1}{8}\psi_{1,0} - \frac{1}{8}\psi_{1,1}.$$

12. As a linear combination of the elements of the basis S_2, ϕ is written as

$$\phi = \phi_{2,0} + \phi_{2,1} + \phi_{2,2} + \phi_{2,3}.$$

17. As a linear combination of the elements of the basis S_3, $\phi_{2,0}$ is written as

$$\phi_{2,0} = \phi_{3,0} + \phi_{3,1}.$$

Compare this to the decomposition of ϕ as a linear combination of the elements of the basis S_1.

24. (a) The projection of $\phi_{1,1}$ onto V_0 is $\frac{1}{2}\phi$.

27. As a linear combination of the elements of the basis B_3, $\phi_{2,0}$ is written as

$$\phi_{2,0} = \frac{1}{4}\phi + \frac{1}{4}\psi + \frac{1}{2}\psi_{1,0}.$$

34. The matrix M_4 is the inverse of A_2.

35. This is a two step process. In the first step, we compute averages and differences to obtain $[-1, 13, 11, -6]$ in $V_1 \oplus V_1^{\perp}$. We again calculate averages and differences on the signal in V_1 to obtain $x_1 = 6$, $x_2 = -7$, $x_3 = 11$, and $x_4 = -6$ in $V_0 \oplus V_0^{\perp} \oplus V_1^{\perp}$. Note that x_1 is the average of all the entries in the signal.

38. The wavelet coefficients are 39.5, 2.5, 22, 9, 16, -16, -18, and -22.

41. (a) The projection of $\sin(t)$ onto V_2 is

$$(1 - \cos(1))\phi(t) + \left(-2\cos\left(\frac{1}{2}\right) + 1 + \cos(1)\right)\psi(t)$$

$$+ \left(-4\cos\left(\frac{1}{4}\right) + 2\cos\left(\frac{1}{2}\right) + 2\right)\psi_{1,0}(t)$$

$$+ \left(2\cos\left(\frac{1}{2}\right) + 2\cos(1) - 4\cos\left(\frac{3}{4}\right)\right)\psi_{1,1}(t).$$

(b) The projection of t^2 onto V_2 is

$$\frac{1}{3}\phi(t) - \frac{1}{4}\psi(t) - \frac{1}{16}\psi_{1,0}(t) - \frac{3}{16}\psi_{1,1}(t).$$

(c) The projection of e^t onto V_2 is

$$(e - 1)\phi(t) + (2\sqrt{e} - 1 - e)\psi(t)$$
$$+ (4\sqrt[4]{e} - 2\sqrt{e} - 2)\psi_{1,0}(t) + (-2\sqrt{e} - 2e + 4\sqrt[4]{e^3})\psi_{1,1}(t).$$

(d) The projection of $\sqrt{t}$ onto V_2 is

$$\frac{2}{3}\phi(t) + \left(\frac{1}{3}\sqrt{2} - \frac{2}{3}\right)\psi(t)$$

$$+ \left(\frac{1}{3} - \frac{1}{3}\sqrt{2}\right)\psi_{1,0}(t) + \left(\sqrt{3} - \frac{1}{3}\sqrt{2} - \frac{4}{3}\right)\psi_{1,1}(t).$$

Chapter 3

3. (b) Since $\|f\| = \sqrt{2}$, f is in $L^2(\mathbb{R})$.

10. The norm of the Mexican hat mother wavelet is $\frac{3}{4}\sqrt{\pi}$.

12. The distance between f and its projection onto V_2 is approximately 0.001895.

20. If we begin with $t_0 = 1$, successive approximations will be $t_1 = 0.5403023059$, $t_2 = 0.857553$, $t_3 = 0.654289$, etc. These approximations converge (slowly) to the solution 0.739085.

21. Starting again with $t_0 = 1$ leads to $t_1 = 1.36788$, $t_2 = 1.25465$, etc. The sequence converges to 1.27846.

37. The function values for the cubic Battle-Lemarié scaling function are

$$\phi(2) = 4\phi(1) = 4\phi(3).$$

67. The limit frequency is $m = \frac{7}{2\pi}$.

Appendix A

4. The set does span $\mathbb{R}^2$.

5. Since $[-1,1] = -[1,-1]$, these two vectors only span a line in $\mathbb{R}^2$.

7. The dimension is 2.

9. Since W is not closed under addition, W is not a subspace of $\mathbb{R}^3$.

11. The angle between the vectors $[1,0,1,1]$ and $[0,1,1,1]$ in $\mathbb{R}^4$ is arccos $\frac{2}{3}$.

12. The angle between $[1,0]$ and $[0,1]$ is 0. This is to be expected because $[1,0]$ and $[0,1]$ are perpendicular vectors.

15. The coefficients of the vector $[1,2,3]$ with respect to the basis B are 2, 2, and 1.

17. The projection $\mathbf{v}_B$ is $\left[\frac{3}{4}, \frac{1}{4}, \frac{7}{4}, \frac{9}{4}\right]$.

18. The projection g_B is $\frac{9}{10}t - \frac{1}{5}$.

Appendix D
Glossary of Symbols

The page number indicates the first page on which the symbol appears.

- ϕ, the father wavelet, page 5

- ψ, the mother wavelet, page 5

- $\psi_{n,k}(t) = \psi(2^n t - k)$, the wavelet daughters, page 9

- $\phi_{n,k}(t) = \phi(2^n t - k)$, the wavelet sons, page 27

- A_n, a wavelet processing matrix, page 12

- $M_n = A_m^{-1}$ (where $m = 2^n$), a wavelet processing matrix, page 34

- B_n, a basis for $\mathbb{R}^{2^n}$ consisting of father, mother, and daughter wavelets, page 12

- S_n, a basis for $\mathbb{R}^{2^n}$ consisting of father and son wavelets, page 28

- C_n, a basis for the orthogonal complement of V_n in V_{n+1}, page 31

- $\langle,\rangle$, an inner product, page 24

- V_n

    - the space of all functions that are piecewise constant on intervals of length $1/2^n$ in $[0,1]$, page 9
    - the space of all functions that are piecewise constant with breaks at rational points of the form $m/2^n$ in $\mathbb{R}$, page 43
    - a subspace of $L^2(\mathbb{R})$ that is a part of a multiresolution analysis, page 53

- $W^\perp$, the orthogonal complement of W, page 25

- c_k, refinement coefficients, page 28

- g_k, high pass filter coefficients, page 63

- h_k, low pass filter coefficients, page 63

- $\mathbf{s}$, a generic signal, page 15

- $\mathbf{v}^T$, the transpose of the vector $\mathbf{v}$, page 4

- $\mathbf{v}_B$, the projection of the vector $\mathbf{v}$ onto the space W with basis B, page 102

- D_4, Daubechies wavelet with four refinement coefficients, page 68

- G, high pass operator, page 71

- H, low pass operator, page 71

- G^*, the dual of the high pass operator, page 74

- H^*, the dual of the low pass operator, page 74

- $L^2([0,1])$, the space of all functions whose squares are integrable on $[0,1]$, page 25

- $L^2(\mathbb{R})$, the space of all functions whose squares are integrable on $\mathbb{R}$, page 44

- $\hat{s}$, the Fourier transform of s, page 79

References

1. Aboufadel, Edward F. 1994. Applications Über Alles: Mathematics for the liberal arts. *PRIMUS*, **4**, no. 4 (December): 317–336.

2. AIMS (Activities Integrating Math, Science, and Technology). 1997. Fingerprinting.
(available at `www.aimsedu.org/Activities/GimmeFive/gimme2.html`)

3. Bradley, Jonathan N., Brislawn, Christopher M., and Hopper, Tom. 1993. The FBI wavelet/scalar quantization standard for gray-scale fingerprint image compression. In *Visual Information Processing II*. Vol. 1961 of *Proc. SPIE*, Orlando, Florida.: SPIE, pp. 293–304.

4. Brislawn, Christopher M. 1995. Fingerprints go digital. *Notices of the A. M. S.*, **42** , No. 11 (November): 1278–1283.

5. Burden, Richard L., and Faires, J. Douglas. 1993. *Numerical Analysis*. 5ed. Boston: PWS-Kent Publishing.

6. Burrus, C. Sidney, Gopinath, Ramesh A., and Guo, Haitao. 1998. *Introduction to Wavelets and Wavelet Transforms: A Primer*. Upper Saddle River, New Jersey: Prentice-Hall.

7. Cipra, Barry. 1993. Wavelet applications come to the fore. *SIAM News*, **27**, No. 7 (November).
(available at `www.siam.org/siamnews/mtc/mtc1193.htm`)

8. Coupland, Douglas. 1994. Toys that bind. *The New Republic*, 6 June: pp. 9–10.

9. Courtney, Steven. 1993. Information age multiplies uses of mathematics formulas for the future. *The Hartford Courant*, 21 October: p. E1.

10. Daubechies, Ingrid. 1992. *Ten Lectures on Wavelets*. Philadelphia: SIAM.

11. Donoho, D. 1993. Nonlinear wavelet methods for recovery of signals, densities, and spectra from indirect and noisy data. In *Different Perspectives on Wavelets, Proceeding of Symposia in Applied Mathematics*, vol. 47. Edited by Ingrid Daubechies. Providence, Rhode Island: American Mathematical Society, pp. 173–205.

12. Daubechies, Ingrid and Lagarias, J.C. 1991. Two scale difference equations I: existence and global regularity of solutions. *SIAM Math. Anal.*, **22**, No. 5: 1388–1410.

13. Federal Bureau of Invesigation. 1999. Integrated automated fingerprint identification system.
(available at www.fbi.gov/programs/iafis/iafis.htm)

14. Foufoula-Georgiou, Efi and Praveen Kumar, eds. 1994. *Wavelets in Geophysics, Wavelet Analysis and Its Applications*, vol. 4. San Diego: Academic Press.

15. Friedman, Avner, and Lavery, John. 1993. *How to Start an Industrial Mathematics Program in the University*. Philadelphia: SIAM.

16. Graps, Amara. 1995. An introduction to wavelets. *IEEE Compu. Sci. Eng.*, **2**, no. 2 (Summer).
(Available at www.amara.com/IEEEwave/IEEEwavelet.html)

17. Greenblatt, Seth A. 1994. Wavelets in econometrics. *The Maple Technical Newsletter*: 10–16.

18. Haar, Alfred. 1910. Zur Theorie der orthogonalen Funktionensysteme. *Mathematische Annalen*, **69**: 331–371.

19. Hubbard, Barbara B. 1996. *The World According to Wavelets*. Wellesley, Massachusetts: A. K. Peters, Inc.

20. Jawerth, Björn and Sweldens, Wim. 1994. An overview of wavelet based multiresolution analyses. *SIAM Review*, **36**, No. 3 (September): 377–412.

21. Kaiser, Gerald. 1994. *A Friendly Guide to Wavelets*. Boston: Birkhäuser.

22. Kobayashi, Mel. 1995. The 'Ueburetto-boom': wavelets in Japan. *SIAM News*, (November), p. 24.

23. Kobayashi, Mel. 1996. Listening for defects: wavelet-based acoustical signal processing in Japan. *SIAM News*, (March), p. 24.

24. Kobayashi, Mel. 1996. Netting namazu: earthquakes and Japanese wavelet research. *SIAM News*, (July/August), p. 20.

25. Meyer, Yves. 1993. *Wavelets: Algorithms and Applications*. Philadelphia: SIAM.

26. Mulcahy, Colm. 1996. Plotting and scheming with wavelets. *Mathematics Magazine*, **69**, No. 5, (December): 323–343.

27. Müller, Peter and Vidakovic, Brani. 1994. Wavelets for kids: a tutorial introduction.
(available at `ftp://ftp.isds.duke.edu/pub/Users/brani/papers/`)

28. Nguyen, Truong and Strang, Gilbert. 1996. *Wavelets and Filter Banks*. Wellesley, Massachusetts: Wellesley-Cambridge Press.

29. Pohlmann, Ken. 1989. *The Compact Disc: A Handbook of Theory and Use*. Madison, Wisconsin: A-R Editions.

30. Rao, Raghuveer M., and Bopardikar, Ajit S. 1998. *Wavelet Transforms: Introduction to Theory and Applications*. Reading, Massachusetts: Addison-Wesley.

31. Schröder, Peter. 1995. Wavelet image compression: beating the bandwidth bottleneck. *Wired*. (May): p. 78.

32. Steen, Lynn Arthur, et. al. 1990. Challenges for college mathematics: an agenda for the next decade. *M.A.A. Focus*, (November).

33. Strang, Gilbert. 1993. Wavelet transforms versus Fourier transforms. *Bulletin of the A.M.S.*, **28**, no. 2 (April): 288–305.

34. Strang, Gilbert. 1989. Wavelets and dilation equations: a brief introduction. *SIAM Rev.*, **31**, no. 4 (December): 614–627.

35. Strichartz, Robert S. 1993. How to make wavelets. *Amer. Math. Month.*, **100**, no. 6 (June-July): 539–556.

36. Stinson, Douglas R. 1995. *Cryptography: Theory and Practice*. Boca Raton: CRC Press.

37. Stollnitz, Eric J., DeRose, Tony D., and Salesin, David H. 1996. *Wavelets for Computer Graphics*. San Frascisco: Morgan Kaufmann.

38. Sweldens, Wim. 1997. Wavelet cascade applet: mathematical background.
(available at `cm.bell-labs.com/cm/ms/who/wim/cascade/math.html`)

39. Sweldens, Wim, and Schröder, Peter. 1996. Build your own wavelets at home.
(available at `cm.bell-labs.com/cm/ms/who/wim/papers/athome.ps.gz`)

40. Von Beyer, Hans C. 1995. Wave of the future. *Discover*, (May): 69–74.

41. Walter, Gilbert G. 1993. Wavelets: a new tool in applied mathematics. *UMAP Journal*, **14**, no. 2 (Summer): 155–178.

42. Walter, Gilbert G.. 1994. *Wavelets and Other Orthogonal Systems with Applications*, Boca Raton: CRC Press.

43. Wang, James Z., et. al. 1997. Wavelet-based image indexing techniques with partial sketch retrieval capability. In *Proceedings of the Fourth Forum on Research and Technology Advances in Digital Libraries*. Los Alamitos, California: IEEE Computer Society.
(available at `www-db.stanford.edu/~wangz/project/imsearch/ADL97/`)

Index

Signal processing, 9, 11–12, 14, 16–19, 32, 35, 37, 39, 71–81, 92, 108–109
Signal
  length, 93
Signals, 4, 70–77
Sons, 27, 31–32, 44
Span, 95
Spanning set, 95
Spline
  B, 60, 88–89
  bell-shaped, 60, 88–89
  linear, 56
Splines
  boundary condition, 88
  interpolation condition, 88
  regularity condition, 88
Strang (Gil), vi, 61, 69
Subspace, 97
Subspace Theorem, 97
Support, 7

T

Thresholding, 12, 18, 20
  hard, 14
  keep or kill, 14
  quantile, 14, 86
  soft, 14
Translations, 49

Two-dimensional wavelet transform, 41
Two-scale difference equation, 54

V

Vector space, 93, 95–96
  n-dimensional real, 94
  dimension, 96
  finite dimensional, 96
Vector
  length, 100
  norm, 99
  projection onto a subspace, 104
Vectors, 93
  angle between, 98, 100
  linearly independent, 96
  orthogonal, 100

W

Wavelet, 6
Wavelet coefficients, 10, 34–35, 41, 86
Wavelet family
  daughters—see Daughters, 6
  father—see Scaling function, 5
  mother—see Mother wavelet, 6
  sons—see Sons, 27
Wavelet scalar quantization standard, 18, 20
World wide web sites, viii, 20, 113, 119–121